Hospitality and Tourism Marketing

Hospitality and Tourism Marketing: Building Customer Driven Hospitality and Tourism Organizations is an absolutely crucial book in light of the world post COVID-19. Following the pandemic, big hotel chains like the Accor Group and Marriot closed outlets, but post COVID-19, the global hospitality and tourism sector is bouncing back.

In bouncing back though, the pandemic brought to the fore the absolute need for high levels of customer centricity in a world that was gripped by fear. This new Hospitality and Tourism (H&T) Marketing book takes a customer-oriented approach to discussing marketing discipline in the global H&T sector. With chapters spanning topics like service characteristics of hospitality and tourism marketing, people management strategies for service businesses, consumer behaviour, designing and building brands, electronic marketing, internet marketing, database marketing, and direct marketing and relationship marketing, this book has the right balance of technology and consumer-oriented topics to provide the right balance for tourism marketing practitioners post pandemic.

Hospitality and Tourism Marketing

Building Customer Driven Hospitality and Tourism Organizations

Robert E. Hinson, Ishmael Mensah,
George Kofi Amoako, Esi Akyere Mensah,
Isaac Sewornu Coffie and Eddy Khosa

Routledge
Taylor & Francis Group
A PRODUCTIVITY PRESS BOOK

First published 2024
by Routledge
605 Third Avenue, New York, NY 10158

and by Routledge
4 Park Square, Milton Park, Abingdon, Oxon, OX14 4RN

Routledge is an imprint of the Taylor & Francis Group, an informa business

ISBN: 978-1-032-68848-0 (hbk)
ISBN: 978-1-032-68847-3 (pbk)
ISBN: 978-1-032-68849-7 (ebk)

DOI: 10.4324/9781032688497

Typeset in Garamond
by Deanta Global Publishing Services, Chennai, India

Contents

Preface

The tourism sector is one of the largest industries in the world. It is one of the fastest-growing industries in many countries today. In 2010, tourism contributed about US$5,992 billion to the global GDP, constituting 9 percent of the global GDP. Though the tourism industry is one of the industries that was most affected by the COVID-19 pandemic, recent results point to progress towards recovery. The contribution of tourism to global GDP before the COVID-19 pandemic was significant. In 2019, the industry's contribution to global GDP was US$9.2 trillion, which represented 10.4 percent of global GDP. The COVID-19 pandemic resulted in a significant decrease in the industry's contribution to global GDP which dropped by 49.1 percent in 2020, representing a loss of US$4.5 trillion and a loss of 62 million jobs worldwide in the same year.

In 2021, tourism's contribution to GDP dropped further to 6.1 percent, with the World Travel & Tourism Council (WTTC) projecting that between 2022 and 2032, tourism's contribution to the global economy is expected to grow at an average annual rate of 5.8% which is more than double the 2.7 percent average annual growth rate estimated for the global economy. Tourism has huge potential for improving the economic fortunes of several countries globally and with artificial intelligence tools, the potential is further amplified.

The hospitality industry which is an essential component of the tourism industry can be defined as encompassing fields within the service industry that include services such as lodging, foodservice, event planning, theme parks, transportation, cruise line, and additional fields within the tourism industry. The industry presents significant opportunities for economic development, particularly in Africa. For instance, according to an article on the Government of Ghana website titled "Jumia Travel launches Ghana Hospitality Industry Review Report for 2016," the report highlights that the country generated GHC 3.7 million (US$ 880,000) in visitor exports in 2015. Additionally, in 2016, Ghana welcomed 1,316,000 international tourist arrivals. The projections outlined in the report suggest that by 2026, there will be a total of 1,913,000 international tourist arrivals, contributing about GHC 4.7 million (US$ 1.12 million) to the Ghanaian economy, reflecting an annual increase of 2.5 percent. The report also underscores the importance of visitor exports as a key component of the direct contribution of Travel & Tourism and that in 2016, Ghana generated GHC 3.7 M (US$ 900,000) in visitor exports which is expected to grow by 5.3 percent in 2017 with 1.3 million international tourist arrivals.

Given the obvious importance of the hospitality and tourism sector to the achievement of several of the United Nation's Millennium Development Goals (Saffu et al., 2008), this book on hospitality and tourism marketing focuses on discussing key pathways to building the marketing and service delivery capacities of hospitality and tourism-related organisations. The book focuses on topics like tourism product management, branding in tourism, understanding the hospitality and tourism marketing process, service characteristics of tourism businesses, electronic marketing, product pricing, and new product development and innovation management. It sets out in very clear terms ways by which marketing can be used as a value and profit driver of tourism organisations the world over.

After reading this book, we expect you to be able to comprehend what tourism and hospitality marketing is, develop service and relational strategies to capitalise on opportunities in the sector, and be ably equipped to develop an appropriate tourism and hospitality marketing mix.

Author Profiles

Professor Robert E. Hinson is currently the Pro-Vice Chancellor at the Ghana Communication Technology, having served previously as the Deputy Vice Chancellor—Academic at the University of Kigali with additional responsibility as Interim Vice Chancellor of the same university. His main research interests lie in the academic areas of marketing and communications, information and technology management, service management, and social responsibility and sustainability management. He has 33 monographs/edited volumes and over 150 peer-reviewed journal papers/book chapters to his credit. He was ranked by the 2021 and 2022 Alper-Doger (AD) Scientific Index as the #1 African Marketing Scholar and leading business and management scholar in Ghana. He is also affiliated with the University of Johannesburg.

Professor Ishmael Mensah is Professor of Tourism and Hospitality Management and Ghana Director of the Confucius Institute at the University of Cape Coast. He holds a PhD Tourism degree from the same university and is Certified Hospitality Educator (CHE) by the American Hotel and Lodging Association and Member of the Institute of Hospitality (MIH). Mensah also holds post-graduate certificates in Hospitality Administration and Event Planning & Tradeshow Management from Georgia State University where he was a fellow under the Ghana Tourism Capacity Development Initiative. He has published widely in high-ranking academic journals in the areas of service quality management, destination marketing, environmental management in hotels, special events management, and community-based tourism development. He is currently the Editor-In-Chief of the *African Journal of Hospitality and Tourism Management.*

Professor George Kofi Amoako is Associate Professor and the Director of Research at the Ghana Communication Technology University.

He obtained his PhD at the London Metropolitan University in the UK and has considerable research, teaching, consulting, and practice experience in the application of marketing theory and principles to everyday marketing challenges and management and organisational issues. George has published extensively in internationally peer-reviewed academic journals (A-, B-, and C-ranked journals) and presented many papers at international conferences in Africa, America, Europe, and Australia. George Amoako has supervised to successful completion six (6) Doctor of Business Administration graduates and is currently supervising five (5) PhD students. He is a PhD examiner for Kwazulu Natal University in South Africa and Open University Malaysia.

Esi Akyere Mensah is Senior Lecturer in Tourism Management with over 9 years of teaching and research experience in Ghanaian, British, and Ivorian institutions. She has taught tourism and business management and sustainability courses at both undergraduate and postgraduate levels. Her research areas cover a wide range of topics, including post COVID-19 tourism recovery, sustainability, and leadership in higher education. She is interested in the intersections of migration, development, and tourism. Her research in tourism focuses on niche tourism such volunteer tourism and host guest interactions in tourism. Her papers have been published in *Tourism Management Perspectives, Journal of International Migration and Integration*, and *International Journal of Tourism Cities and Anatolia.* She is currently an editorial board member of the *Tourism Planning and Development Journal*, a copyeditor, and a reviewer for tourism journals and corporate institutions.

Isaac Sewornu Coffie is Part-Time Lecturer at the Accra Technical University and holds a PhD in Marketing from the Department of Marketing and Entrepreneurship at the University of Ghana Business School. His teaching and research areas are social marketing, corporate social responsibility (CSR), entrepreneurship, and market orientation. He has over 12 years of corporate working experience as a marketing executive. He is also an entrepreneur. He is published in the *Social Marketing Quarterly, Journal of Social Marketing*, and *International Review on Public and Nonprofit Marketing.*

Eddy Khosa is Former CEO of the Johannesburg Tourism Company, Former Interim Chairperson of the CATHSSETA Board—Culture, Arts, Tourism, and

Hospitality SETA, and Former General Manager of AH Hotel & Conferences in Accra—Ghana. He holds a Master of Science (MSc) in Global Management Degree from the University of Salford, Manchester, UK.

Chapter 1

Understanding Marketing

Chapter Outline

- Introduction to Marketing
- Competing Business Orientations
- Components of Societal Marketing Concept
- Socially Responsive Marketing and the Triple Bottom Line

Chapter Objectives

At the end of this chapter, you should be able to:

- Define marketing.
- Explain the production concept.
- Explain the product concept.
- Explain the selling concept.
- Explain the marketing concept.
- Explain the societal marketing concept.

Introduction

This chapter introduces you to the fundamentals and concepts of marketing. First, the chapter provides and discusses various concepts in marketing, then highlights some challenges that arise out of the practice of tourism

DOI: 10.4324/9781032688497-1

marketing. The chapter further provides a discussion on five marketing philosophies and highlights other marketing-related concepts such as governance, leadership, business planning, and culture. This book will equip the readers with knowledge on the role of marketing in the tourism and hospitality industry—from planning through to execution and evaluation.

Introduction to Marketing

The concept of marketing has been defined in different ways. The Chartered Institute of Marketing (UK) defines it as the management process of identifying, anticipating, and satisfying customer requirements profitably; whilst the American Marketing Association defines it as the process of planning and executing, conception, pricing, promotion, and distribution of ideas, goods, and services to create exchanges that satisfy individual and organisational objectives. From these definitions it could be deduced that marketing is a process that culminates in satisfied consumers and profitable organisations.

Kotler also defines marketing as:

> the science and art of exploring, creating, and delivering value to satisfy the needs of a target market at a profit. Marketing identifies unfulfilled needs and desires. It defines, measures, and quantifies the size of the identified market and the profit potential. It pinpoints which segments the company is capable of serving best and it designs and promotes the appropriate products and services.

Marketing is also the process of communicating the value of a product or service to customers. Marketing might sometimes be interpreted as the art of selling products, but selling is only a small fraction of marketing. Traditionally, marketing is how an organisation communicates to, connects with, and engages its target audience to convey the value of and ultimately sell its products and services. However, since the emergence of digital media, in particular social media and technology innovations, it has increasingly become more about companies building deeper, more meaningful, and lasting relationships with the people who want to buy their products and services. The ever-increasingly fragmented world of media complicates marketers' ability to connect and, at the same, time presents incredible opportunities to penetrate global markets through e-marketing.

From a societal point of view, marketing is the link between a society's material requirements and its economic patterns of response. Marketing satisfies these needs and wants through exchange processes and building long-term relationships. Marketing can be looked at as an organisational function and a set of processes for creating, delivering, and communicating value to customers, and managing customer relationships in ways that benefit the organisation and its shareholders. It is the science of choosing target markets through market analysis and market segmentation, as well as understanding consumer buying behaviour and providing superior customer value.

The Five Competing Business Orientations

There are five competing concepts under which organisations can choose to operate their business. These are the production concept, the product concept, the selling concept, the marketing concept, and the societal marketing concept. The societal marketing concept comprises of four components, namely, relationship marketing, internal marketing, integrated marketing, and socially responsive marketing. The set of engagements necessary for successful marketing management includes capturing marketing insights, connecting with customers, building strong brands, shaping the market offerings, delivering and communicating value, creating long-term growth, and developing marketing strategies and plans.

Marketing management is the conscious effort to achieve the desired exchange outcomes with target markets. Organizations undertake marketing activities based on certain philosophies. A marketing philosophy is a basic idea that underpins all activities related to the marketing of the products or services of an organisation. But what philosophy should guide a company's marketing efforts? What relative weights should be given to the often-conflicting interests of the organisation, customers, and society? Clearly, marketing activities should be carried out under a well-thought-out philosophy of efficient, effective, and socially responsible marketing. The five competing concepts under which organisations conduct marketing activities are production concept, product concept, selling concept, marketing concept, and societal marketing concept.

The Production Concept

The production concept, one of the oldest in business, holds that consumers prefer products that are widely available and inexpensive. Managers of production-oriented businesses concentrate on achieving high production efficiency, low costs, and mass distribution. This orientation makes sense in developing countries, where consumers are more interested in obtaining the product than in its features. It is also used when a company wants to expand its market. Texas Instruments is a leading exponent of this concept. It concentrates on building production volume and upgrading technology to bring costs down, leading to lower prices and expansion of the market. This orientation has also been a key strategy for many Japanese companies.

The Product Concept

Other businesses are guided by the product concept, which holds that consumers favour those products that offer the most quality, performance, or innovative features. Managers in these organisations focus on making superior products and improving them over time, assuming that buyers can appraise quality and performance. Product-oriented companies often design their products with little or no customer input, trusting that their engineers can design exceptional products. A General Motors executive said years ago: "How can the public know what kind of car they want until they see what is available?" GM today asks customers what they value in a car and includes marketing in the very beginning stages of design. The product concept can lead to marketing myopia. Marketing myopia which was first used in an article by Theodore Levitt in the *Harvard Business Review*, refers to a short-sighted and inward approach to marketing that focuses on the needs of the organisation rather than that of the customer.

The Selling Concept

The selling concept, another common marketing orientation, holds that consumers and businesses if left alone, will ordinarily not buy enough of the organisation's products. The organisation must, therefore, undertake an aggressive selling and promotion effort. This concept assumes that consumers must be coaxed or forced into buying, so the company has a battery of selling and promotion tools to stimulate buying. The selling concept is practised most aggressively with unsought goods—goods that buyers normally do not think of buying, such as insurance and funeral policies. The selling concept is also practised in the non-profit area by fundraisers, college admissions offices, and political parties. Most firms practice the selling concept when they have overcapacity. They aim to sell what they make rather than make what the market wants. In modern industrial economies, productive capacity has been built up to a point where most markets are buyer markets (the buyers are dominant) and sellers have to scramble for customers. Prospects are bombarded with sales messages. As a result, the public often identifies marketing with hard selling and advertising. But marketing based on hard selling carries high risks. It assumes that customers who are coaxed into buying a product will like it; and if they don't, they won't bad-mouth it or complain to consumer organisations and will forget their disappointment and buy it again. These are indefensible assumptions. In fact, one study showed that dissatisfied customers may bad-mouth the product to ten or more acquaintances; bad news travels fast, something marketers who use hard selling should bear in mind.

The Marketing Concept

The marketing philosophy is a concept that is based on central tenets crystallised in the mid-1950s. It challenges the three previous marketing orientations. The marketing concept holds that the key to achieving organisational goals consists of the company being more effective than its competitors in creating, communicating, and delivering value to its chosen target markets.

Theodore Levitt of Harvard drew a perceptive contrast between the selling and marketing concepts:

> Selling focuses on the needs of the seller; marketing focuses on the needs of the buyer. Selling is preoccupied with the seller's need to convert his/her product into cash; marketing is preoccupied with the idea of satisfying the needs of the customer through the product and the whole cluster of things associated with creating, delivering, and finally consuming it.

The marketing concept rests on four pillars: target market, customer needs, integrated marketing, and profitability. The selling concept takes an inside–out perspective. It starts with the factory, focuses on existing products, and calls for heavy selling and promotion to produce profitable sales. The marketing concept takes an outside–in perspective. It starts with a well-defined market, focuses on customer needs, coordinates activities that affect customers, and produces profits by satisfying customers. Companies do best when they choose their target market(s) carefully and prepare tailored marketing programmes.

The Societal Marketing Concept

The societal marketing concept calls upon marketers to build social and ethical considerations into their marketing practices. They must balance and juggle the often-conflicting criteria of company profits, customer satisfaction, and public interest. Yet several companies have achieved notable sales and profits by adopting and practicing the societal marketing concept. These include the PLANET 21 initiative by Accor Hotels geared towards ensuring environmental sustainability, and Omni Hotels' "Say Goodnight to Hunger" programme. Through this the hotel has donated more than 3 million meals to needy American families. Societal marketing creates a favourable image for an organisation or destination and ultimately helps increase sales. As was alluded to earlier, the four components of societal marketing are relationship marketing, internal marketing, integrated marketing, and socially responsive marketing and these will now be discussed in turn.

Relationship Marketing (RM)

Marketing is no longer simply about developing, selling, and delivering products. It is also progressively more concerned with the development and maintenance of mutually satisfying long-term relationships with customers. Enduring relationships with customers cannot be duplicated by competitors, and therefore provide a unique and sustained competitive advantage. The expression most widely used to describe this new form of marketing is relationship marketing (RM). Other terms have been used, either as substitutes for RM or to describe some close parallel – micromarketing, database marketing, one-to-one marketing, loyalty marketing, wrap-around marketing, customer partnering, symbiotic marketing, and interactive marketing. Successful relationship marketing is often characterized by the establishment, maintenance, and enhancement of relationships with customers, and other partners, at a profit, so that the objectives of the parties involved are met. This is achieved by a mutual exchange and fulfilment of promises.

Internal Marketing

According to Kotler and Keller (2012), internal marketing is the task of recruiting, developing, and motivating employees to ably serve the customer. They contend that internal marketing requires everybody in the organisation to buy into the concepts and goals of marketing and to choose to provide and communicate customer value. They argue that it is only when all the employees in the organisation realise that their job is to "create, serve, and satisfy customers does the company become an effective marketer" (Kotler and Keller, 2012). Internal marketing, according to Kotler and Keller (2012), must take place at two levels. They note that at the first level, various marketing functions must work together, while at the second level, other departments must embrace and think "marketing." They concede that internal marketing "requires vertical alignment with senior management and horizontal alignment with other departments, so everyone understands, appreciates, and supports the marketing effort" (Kotler and Keller, 2012, p64;).

Internal marketing (IM) as a concept was developed in the 1980s. Although IM has been conceptualised and defined variously by different writers, it is commonly agreed that successfully pursuing an internal marketing programme is critical to delivering superior quality service and successful external marketing (Berry and Parasuraman, 1991). Primarily, the focus of internal marketing is on employees whose function is to provide services to both internal and external customers.

Research has shown that effective internal marketing has positive impact on corporate brand image and enables employees to work to their optimum best (Papasolomou and Vrontis, 2006; Alhakimi and Alhariry (2014) . IM enables employees to develop and enhance a service culture and an internal environment that allows them to give the best possible customer treatment (Gronroos, 1990; Caruana and Calleya, 1998). Scholars (e.g., Gronroos, 1990; Rafiq and Ahmed, 1993) have argued that the ultimate objective of internal marketing is to develop a service culture that will help to establish customer consciousness among employees. The ultimate goal of internal marketing, according to Berry and Parasuraman (1991, p151) "is to encourage effective marketing behaviour; build an organisation of marketers willing and able to create customers for the firm." The authors note that "the ultimate strategy of internal marketing is to create true customers of employees" (ibid). They observed that internal marketing means that organisations must treat their employees as they expect their employees to treat their customers.

Integrated Marketing

Integrated marketing refers to an approach to create a unified and seamless experience for consumers to interact with the tourism brand/enterprise (e.g., the Fiesta Royale Hotel). Integrated marketing welds all aspects of marketing communication such as advertising, sales promotion, public relations, direct marketing, and social media, through their respective mix of tactics, methods, channels, media, and activities, so that all work together as a unified force. The American Marketing Association also argues that integrated marketing is a planning process designed to assure that all brand contacts received by a customer or prospect for a product, service, or organisation are relevant to that person and consistent over time.

Societal Responsive Marketing

Societally responsive (or socially responsible) marketing refers to the idea that businesses should balance profit-making activities with activities that benefit society. Societally responsive marketing can be practised by organisations through the use of corporate social initiatives like cause-related marketing. According to Laczniak and Murphy (2012),societal responsive marketing refers to a strategic approach in which businesses and marketers actively consider and address the broader societal impact of their marketing activities. It involves a commitment to understanding and responding to the diverse needs, values, and concerns of the larger society, beyond simply meeting customer demands. This approach goes beyond profit-driven motives and aims to contribute positively to social well-being, environmental sustainability, and ethical considerations. Societal responsive marketing encompasses practices such as corporate social responsibility, sustainable business practices, and ethical marketing strategies that align with societal values and expectations. The goal is to build long-term relationships with customers while fostering positive contributions to the community and the world at large. They also believe that customers will increasingly look for demonstrations of good corporate citizenship. Smart companies will respond by adding "higher order" image attributes rather than simply rational and emotional benefits. Critics, however, complain that cause-related marketing might make consumers feel they have fulfilled their philanthropic duties by buying products instead of donating to causes directly.

It is important to stress however that companies that pursue socially responsive marketing should practice it ideally from the standpoint of the triple bottom line (TBL). Designing a triple bottom line sustainability development includes governance, leadership, a business plan, measuring and reporting, organisational learning, culture, and information systems; these will be discussed in turn.

Governance

An effective governance structure is a prerequisite for the triple bottom line's sustainability (Amran et al 2014; Bauer Ho, 2005). According to Bauer et al. (2008) , organizations with committed top management and well-governed firms significantly outperform poorly governed firms by up to 15% a year. Given the global significance of sustainability, stakeholders anticipate companies to conduct their business activities in a sustainable manner to achieve their vision and mission (Tjahjadi et al., 2021; Hussain et al., 2018). The success of this sustainability objective is heavily contingent on the quality of corporate governance. Two prevalent corporate governance systems worldwide are the one-tier system and the two-tier system. In the one-tier system, the Board of Directors (BoD) serves as both supervisors and executors, while in the two-tier system, the BoD takes on the supervisory role, and the top management team becomes the executors. The division of responsibilities in a two-tier system enhances supervision quality and augments transparency in decision-making. Within this framework, the BoD holds the responsibility of overseeing and advising managers on strategic matters and decision-making processes (Hussain et al., 2018). . A key problem is how to ensure accountability whilst promoting the innovation necessary for the development of a strategic rather than a compliance-based perspective on sustainability (Dunphy et al., 2007).

Leadership

Leadership entails examining how senior leaders in organisations are adopting principles of sustainability to enact the traditional tasks of leadership; i.e., setting direction, creating alignment, and maintaining commitment within the context of sustainability (Quinn and Dalton, 2009). To achieve success, leaders pursuing a sustainability agenda must pay attention to how the concept is framed and introduced into the organisation. Such leaders must build capacity in their systems (educational, communication, rewards, performance, etc.), including broad and deep stakeholder engagement. It can be argued that leaders adopting sustainability practices are similar to other "effective" leaders. However, these leaders must also have additional capacity and mindset to include a wider expanse of stakeholders and a different mindset as to the purpose of the organisation.

In addressing how leaders achieve organisational support and integrate sustainability into organisational systems, processes, and structures, Morsing and Oswald (2009) examined how senior executives adopt appropriate management control systems to communicate to employees and other stakeholders. Hind et al. (2009) also identified the basic qualities and skills of leadership for integrating social and sustainability concerns into an organisation. These authors conclude that leaders who take social issues seriously enough to want to mainstream them into their business, question business as usual by being open to new ideas and challenging others to adopt new ways of working, understand the role of each player in society (government, unions, business, social partners, non-governmental organisations, and civil society), and how they interact with each other, and build internal and external partnerships by taking a multidisciplinary approach and creating strategic networks and alliances. They also identify stakeholders, build relations with internal and external stakeholders, engage in dialogues, balance competing demands, respect diversity, and take a strategic view of the business environment.

Business Planning

The third element in the sustainable triple bottom line performance is business planning. Searcy (2009) recommends that any business plan for corporate sustainability should be split into three diagnostics: situational, goal, and implementation. An appropriate course for sustainability management must be outlined at the very beginning of the process with clarity regarding who is responsible for what. Questions such as "What outcomes are desirable?" and "How do the key sustainability priorities conflict with one another, if at all?" must be clearly answered and communicated from the outset. Other important questions address the following: "What are the key roadblocks?" "Who is responsible for collecting, analysing, and reporting the data?"

Measuring and Reporting

In order to develop a successful sustainability programme, an organisation's activities have to be measured against its identified objectives, and data and progress meaningfully reported (Edwards, 2009). In recent years, there has been growing pressure on organisations to measure and report on triple bottom line performance which hitherto had been regarded as outside an organisation's responsibility. The Global Reporting Initiative (GRI), produces one of the world's most used standards for sustainability reporting. The GRI provides a standard detail in a 40-page report, information relevant to the key criteria, and information for judging an organisation's sustainability. However, it has been argued that the current GRI guidelines are not sufficient for assuring that a report answers the questions of how sustainable a company is, or how quickly it is approaching sustainability (Isaksson and Steimle, 2009).

Organisational Learning

Several authors have stressed the importance of organisational learning in the pursuit of sustainability (e.g., Nattrass and Altomare, 1999; Senge et al., 1999), and links between organisational learning and sustainability have shown signs of increasing convergence (Senge and Carstedt, 2001; Molnar and Mulvihill, 2003). Some of the most commonly cited characteristics of a learning organisation include challenging mental models, fostering fundamental change, engaging in extensive collaborative activity, and revisiting core assumptions about business and its purpose (Jamali, 2006). The notions of systems-level thinking and learning culture including fostering a culture of learning and experimentation are critical in organisational learning. Jamali (2006) mentioned that there is a history in organisations of emphasising the exploitation of new ideas without paying equal attention to the more time-intensive process of creative exploration. As a result, organisations have developed a habit of quick fixes. Furthermore, learning at the organisational level involves creating systems/processes, to capture knowledge, support knowledge creation, and empower continuous transformation. In their hunt for sustainable development and TBL integration, organisations must efficiently and effectively create, capture, harvest, shape, and apply sustainability-related knowledge and insights.

Culture

The sixth element in designing a triple bottom line sustainability performance is culture. A number of authors suggest that organisational culture is very important for nurturing sustainability (Lacey et al., 2009; Rimanoczy and Pearson, 2010). Organisational culture refers to the underlying beliefs, assumptions, values, and ways of interacting within the organisation that culminates in a unique socio-psychological environment of an organisation. The culture of an organisation must align with the triple bottom line to foster sustainability.

Information Systems

Finally, information systems (IS) constitute the seventh element in designing a comprehensive TBL sustainability performance programme. However, research on how information systems (IS) can support the development of sustainability in organisations has received little attention. Chen et al. (2008) suggest a conceptual model and propositions concerning the roles of IS in the pursuit of ecological sustainability. Their model advocates that under different institutional pressures, IS can be leveraged to achieve eco-efficiency, eco-equity, and eco-effectiveness through automating, information sharing, and transforming organisations, respectively. They further note that IS professionals must take into consideration the environmental impact of the technologies they design and produce for organisational use.

Activity

1. Explain the term "marketing" and outline the various orientations of marketing in relation to the development of the hospitality and tourism sector.
2. Define and explain two advantages of each of the following concepts in relation to the hospitality and tourism industry.

a. Relationship marketing.
b. Internal marketing.
c. Integrated marketing.

3. As an executive in the hospitality industry, explain the concept of societally responsive marketing and briefly explain how you will successfully implement the concept within the hospitality industry using the triple bottom line (TBL)?

References

Alhakimi, W., & Alhariryb, K. (2014). Internal marketing as a competitive advantage in banking industry. *Academic Journal of Management Sciences ISSN, 2305*, 2864.

Amran, A., Lee, S. P., & Devi, S. S. (2014). The influence of governance structure and strategic corporate social responsibility toward sustainability reporting quality. *Business Strategy and the environment, 23*(4), 217–235.

Bauer, R., Frijns, B., Otten, R., & Tourani-Rad, A. (2008). The impact of corporate governance on corporate performance: Evidence from Japan. *Pacific-Basin Finance Journal, 16*(3), 236–251.

Caruana, A., & Calleya, P. (1998). The effect of internal marketing on organisational commitment among retail bank managers. *International Journal of Bank Marketing, 16*(3), 108–116.

Chen, A. J., Boudreau, M. C., & Watson, R. T. (2008). Information systems and ecological sustainability. *Journal of Systems and Information Technology, 10*(3), 186–201.

Dunphy, D., Griffiths, A., & Benn, S. (2007). *Organizational change for corporate sustainability* (2nd ed.). London: Routledge.

Edwards, M. G. (2009). An integrative metatheory for organizational learning and sustainability in turbulent times. *The Learning Organization, 16*(3), 189–207.

Gronroos, C. (1990). Relationship approach to marketing in service contexts: The marketing and organizational behavior interface. *Journal of Business Research, 20*(1), 3–11.

Hind, P., Wilson, A., & Lenssen, G. (2009). Developing leaders for sustainable business. *Corporate Governance: The International Journal of Business in Society, 9*(1), 7–20.

Ho, C. K. (2005). Corporate governance and corporate competitiveness: An international analysis. *Corporate Governance: An International Review, 13*(2), 211–253.

Hussain, N., Rigoni, U., & Orij, R. P. (2018). Corporate governance and sustainability performance: Analysis of triple bottom line performance. *Journal of business ethics, 149*, 411–432.

Isaksson, R., & Steimle, U. (2009). What does GRI-reporting tell us about corporate sustainability?. *The TQM Journal, 21*(2), 168–181.
Jamali, D. (2006). Insights into triple bottom line integration from a learning organization perspective. *Business Process Management Journal, 12*(6), 809–821.
Kotler, P., & Keller, K. L. (2012). *Marketing management 14th edition*. Pearson Education..
Lacey, T. A., & Wright, B. (2009). Employment outlook: 2008–18-occupational employment projections to 2018. *Monthly Labor Review, 132*, 82.
Laczniak, G. R., & Murphy, P. E. (2012). Stakeholder theory and marketing: Moving from a firm-centric to a societal perspective. *Journal of Public Policy &Marketing, 31*(2), 284–292.
Molnar, E., & Mulvihill, P. R. (2003). Sustainability-focused organizational learning: Recent experiences and new challenges. *Journal of Environmental Planning and Management, 46*(2), 167–176.
Morsing, M., & Oswald, D. (2009). Sustainable leadership: Management control systems and organizational culture in Novo Nordisk A/S. *Corporate Governance: The International Journal of Business in Society, 9*(1), 83–99.
Nattrass, B., & Altomare, M. (1999). *The natural step for business: Wealth, ecology and the evolutionary corporation*. New Society Publishers.
Papasolomou, I., & Vrontis, D. (2006). Building corporate branding through internal marketing: The case of the UK retail bank industry. *Journal of product & brand management, 15*(1), 37–47.
Quinn, L., & Dalton, M. (2009). Leading for sustainability: Implementing the tasks of leadership. *Corporate Governance: The International Journal of Business in Society, 9*(1), 21–38.
Rafiq, M., & Ahmed, P. K. (1993). The scope of internal marketing: Defining the boundary between marketing and human resource management. *Journal of Marketing Management, 9*(3), 219–232.
Rimanoczy, I., & Pearson, T. (2010). Role of HR in the new world of sustainability. *Industrial and Commercial Training, 42*(1), 11–17.
Saffu, K., et al. (2008). The contribution of human capital and resource-based view to small- and medium-sized tourism venture performance in Ghana. *International Journal of Emerging Markets.*
Searcy, C. (2009). Setting a course in corporate sustainability performance measurement. *Measuring Business Excellence, 13*(3), 49–57.
Senge, P. M., Carstedt, G., & Porter, P. L. (2001). Next industrial revolution. *MIT Sloan Management Review, 42*(2), 24–38.
Senge, P., Kleiner, A., Roberts, C., Ross, R., Roth, G., Smith, B., & Guman, E. C. (1999). The dance of change: The challenges to sustaining momentum in learning organizations.
Tjahjadi, B., Soewarno, N., & Mustikaningtiyas, F. (2021). Good corporate governance and corporate sustainability performance in Indonesia: A triple bottom line approach. *Heliyon, 7*(3), 1–23.

Chapter 2

Marketing for Hospitality and Tourism

Chapter Outline

- Introduction
- Hospitality and Tourism Marketing
- Marketing in Hospitality and Tourism Industry
- The Hospitality Industry
- Customer Value and Satisfaction in the Hospitality Industry

Chapter Objectives

- Explain the relationships between the world's hospitality and travel industry.
- Explain the nature of marketing in the hospitality industry.
- Explain the relationships between customer value and satisfaction.
- Explain why the marketing concept calls for customer orientation.
- Describe the concept of the lifetime value of a customer and relate it to customer loyalty and retention.

Introduction

Tourism involves all the activities connected with travel to and stay in destinations outside people's places of residence, while hospitality includes all

DOI: 10.4324/9781032688497-2

facilities created to cater for the needs of visitors such as hotels, restaurants, nightclubs, and theme parks. As a manager in the hospitality and tourism industry, you will need to become familiar with the principles and practices of marketing. Today, the customer is king. Satisfying the customer is a priority in most businesses. Especially within the hospitality and tourism industry which is increasingly becoming competitive. Managers of hospitality and tourism businesses must realise that they cannot satisfy all customers; they have to choose their customers carefully. They must select those customers who will enable the company to meet its objectives. To compete effectively for their chosen customers, companies must create a marketing mix that gives their target markets more value than their competitor's marketing mix.

Hospitality and Tourism Marketing

The hospitality industry is one of the world's largest industries. The industry has evolved from the 1960s till now with decades of improvement within the industry (Dev et al., 2010). However, it is also a very competitive industry as destinations and local and global hospitality and tourism firms jockey for strategic positions within the global tourism industry. Marketing has assumed an increasingly important role in ensuring the competitiveness of tourist destinations and hospitality businesses.

The entrance of corporate giants into the hospitality market has transformed it from a mom-and-pop industry, where individually owned restaurants and hotels were the norms, into an industry dominated by chains. These chains operate in a highly competitive environment where aggressive marketing skills are employed to win customers. The marketing capabilities of these large firms have created a competitive marketing environment. In response to growing competitive pressures, hotel chains rely more on the expertise of a marketing director. The position of food and beverage manager or rooms division manager is no longer the only career path leading to the general manager's position, in many chains, the position of marketing director is emerging as an alternative career path to the general manager. Some hotel chains have created a structure in which the marketing director reports to a corporate manager, thus elevating the hotel's chief marketer to the same level as the general manager. Marketing is a philosophy needed by all managers. While the marketing director is a full-time marketer, everyone else must be a part-time marketer. This would suggest that given the competitive nature of the industry, everybody within the organisation should be marketing oriented.

Marketing in Hospitality and Tourism Industry

The principal concepts of traditional marketing are applicable to hospitality and tourism marketing. However, tourism marketing which falls within the domain of services marketing is a complex phenomenon which could be viewed from both the micro and macro levels. At the macro level, the role of the National Tourism Organisations (NTOs) or National Tourist Boards (NTBs) and Convention and Visitors' Bureaux (CVBs) are crucial for the successful marketing of destinations. For instance, in Ghana, the Ghana Tourist Authority (GTA), as part of its functions is supposed to market tourism locally and internationally. The micro level looks at the roles of various tourism and hospitality businesses such as hotels, restaurants, car rentals, convention centres, shopping centres, transport companies, and other tourist attractions that market their products within the context of the destination as a whole.

A product can be "ideas, goods, or services." Since tourism is primarily a service-based industry, the principal products provided by tourism businesses are recreational experiences and hospitality. In practice, the marketing of services is different from that of physical goods. With regards to tourism, it is an experience made up of many components that are marketed rather than a tangible product. These are intangible products and are more difficult to market than tangible products such as automobiles. Services are different from products because while goods are produced, services are performed. Generally, hospitality and tourism services (HTS) are characterised by intangibility, inseparability, perishability, and heterogeneity (This will be discussed in greater detail in Chapter 3). The intangible nature of services makes quality control difficult but crucial. It also makes it more difficult for potential customers to evaluate and compare service offerings. In addition, instead of moving the product to the customer, the customer must travel to the product (area/community). Travel is a significant portion of the time and money spent in association with recreational and tourism experiences and is a major factor in people's decisions on whether or not to visit a destination.

Marketing hospitality and tourism services are quite complex due to the fact that the industry is multi-sectoral and multidimensional. There are a wide array of actors and businesses that are often fragmented but must collaborate to ensure the successful marketing of the destination. As an industry, tourism has many components comprising the overall "travel experience." Along with transportation, it includes such things as accommodations, food and beverage services, shops, entertainment, aesthetics, and special events. It is rare for one business to provide the variety of activities or facilities tourists need or desire (Hsu et al., 2022). This adds to the difficulty of maintaining and controlling the quality of the experience. To overcome this hurdle, tourism-related businesses, agencies, and organisations need to work together to package and promote tourism opportunities in their areas and align their efforts to assure consistency in product quality.

The marketing strategy, or mix, should be viewed as a package of offerings designed to attract and serve the customer or visitor. Recreation and tourism businesses and communities should develop both external and internal marketing mixes for different target markets. The external marketing mix includes product/service, price, place/location, and promotion.

In the hotel industry, marketing and sales are often thought to be the same, and no wonder the sales department is one of the most visible in a hotel. However, while sales is more transaction-based, marketing seeks to win and retain customers for the long term. Sales managers focus on selling the travel experience such as hotel rooms, tours, and food and beverage to prospective clients. Thus, the sales function is highly visible, whereas most of the non-promotional areas of the marketing function take place behind closed doors. Other managers confuse marketing with advertising and sales promotion. It is not uncommon to hear restaurant managers say that they "do not believe in marketing" when they actually mean that they are disappointed with the impact of their advertising. In reality, selling and advertising are only two marketing functions, and often not the most important. Advertising and sales are components of the promotional element of the marketing mix. Other marketing mix elements include product, price, and distribution. Marketing also includes research, information systems, and planning.

The four-P framework calls upon marketers to decide on the product and its characteristics, set the price, decide how to distribute their product, and choose methods for promoting their product (Jager, 2007; Langat, 2016). For example, McDonald's has a fast-food product. It uses quality ingredients and develops products that can sell at the prices people expect to pay for fast food. Most people will not spend more than fifteen minutes to travel to a McDonald's restaurant. As part of its distribution plan, McDonald's must have restaurants that are conveniently located to its target market. Finally, McDonald's appeals to different market segments by offering different menus and having many units throughout the city. This allows McDonald's to make effective use of mass media, such as television. The marketing mix must be just that—a mix of ingredients to create an effective marketing offer for the target market. In the subsequent unit, we introduce three more "Ps" developed for services. The issue is not whether there should be four, seven, or ten Ps so much as what framework is most helpful in designing a marketing strategy. The marketer sees the four Ps as a cabinet of tools that can guide market planning.

If marketers do a good job of identifying consumer needs, developing a good product, and pricing, distributing, and promoting it effectively, the result will be attractive products and satisfied customers. Marriott developed its Courtyard concept, Deji Akinyanju developed the Chicken Republic, a Nigerian fast-food chain and franchise that specialises in chicken recipes, and Mrs. Fields introduced her cookies, and they were swamped with customers. They designed differentiated products, offering new consumer benefits. Marketing means "hitting the mark." Peter Drucker, a leading management thinker, put it this way: "Marketing aims to make selling superfluous. The aim is to know and understand customers so well that the product or service fits them and sells itself." This does not mean that selling and promotion are unimportant, but rather, they are part of a larger marketing mix, a set of marketing tools that work together to produce satisfied customers. The only way selling and promoting will be effective is if we first define customer targets and then prepare an easily accessible and available value package.

The Hospitality Industry

The hospitality industry is mainly focused on the provision of a broad category of services including lodging, restaurants, event planning, theme parks, transportation, cruise line, and entertainment for travellers during their stay at the destination. The hospitality industry is a several billion-dollar industry that mostly depends on the availability of leisure time and disposable income. A hospitality unit such as a restaurant, a hotel, or an amusement park consists of multiple departments such as facility maintenance, direct operations (servers, housekeepers, porters, kitchen workers, bartenders, etc.), finance, administration, marketing, and human resources.

Tourism

The World Tourism Organisation defines tourists as people "traveling to and staying in places outside their usual environment for not more than one consecutive year for leisure, business, and other purposes." Tourism involves travel to places outside one's usual place of residence for different purposes such as recreation, religion, leisure, or business and the facilities created to cater to the needs of the tourist. Hospitality arises out of the need to provide services and facilities to cater to the needs of the tourist. Thus hospitality is a subset of tourism (Ionel, 2016).

Customer Value and Satisfaction in the Hospitality Industry

Today's marketing is not simply a business function: it's a philosophy, a way of thinking, and a way of structuring your business and your mind. Marketing is more than a new ad campaign or promotional strategy; marketing is part of everyone's job, from the receptionist to the board of directors. The task of marketing is never to fool the customer or endanger the company's image. The marketer's task is to design a product–service combination that provides real value to targeted customers, motivates purchase, and fulfils genuine consumer needs.

Marketing, more than any other business function, deals with customers. Creating customer value and satisfaction are at the heart of the hospitality and travel industry marketing. Many factors contribute to making a business successful, however, today's successful companies at all levels have one thing in common—they are strongly customer focused and heavily committed to marketing. Accor has become one of the world's largest hotel chains by delivering L'espirit Accor, the ability to anticipate and meet the needs of their guests, with genuine attention to detail. Ritz-Carlton promises and delivers truly "memorable experiences" for its hotel guests. McDonald's grew into the world's largest restaurant chain by providing its guests with QSC&V (quality, service, cleanliness, and value). These and other successful hospitality companies know that if they take care of their customers, market share and profits will follow.

Whether you want to be a restaurant manager, executive housekeeper, or any other hospitality career choice, marketing will directly affect your personal and professional life. As a manager, you will be motivating your employees to create superior value for your customers. You will want to make sure that you deliver customer satisfaction at a profit. This is the simplest definition of marketing. This book will start you on a journey that will cause your customers to embrace you and make marketing your management philosophy (Oh & Kim, 2017).

Customer Orientation

The purpose of a business is to create and maintain satisfied, profitable customers. Customers are attracted and retained when their needs are met. Not only do they return to the same cruise line, hotel, rental car firm, and restaurant, but they also talk favourably to others about their satisfaction. Customer satisfaction leading to profit is the central goal of hospitality marketing.

"What about profits?" Hospitality managers sometimes act as if today's profits are primary and customer satisfaction is secondary. This attitude eventually sinks a firm as it finds fewer repeat customers, and faces increasingly negative word of mouth. Successful managers understand that profits are best seen as the result of running a business well rather than as its sole purpose. When a business satisfies its customers, the customers will pay a fair price for the product. A fair price includes a profit for the firm. Managers who forever try to maximise short-run profits are short-selling both the customer and the company.

A customer arrived at a restaurant before closing time and was greeted with "What do you want?" Somewhat surprised, the customer replied that he would like to get a bite to eat. A surly voice informed the customer that the restaurant was closed. At this point, the customer pointed to a sign on the door stating that the restaurant was open until 9:00 p.m. "Yeah, but by the time I clean up and put the food away it'll be nine, so we're closed." The customer left and went to another restaurant a block away and never returned to the first restaurant.

Let's speculate for a moment. Why was the customer treated in such a shabby manner? Perhaps:

The employee wanted to leave early.

The employee suffered from a headache.

The employee had personal or family problems.

What really happened in the restaurant episode is that this employee once served a customer right before closing time, resulting in the employee working until 10:30 p.m. Instead of the corporate office thanking her for serving the customer and staying late, it reprimanded her for putting in extra time. The corporate office wanted to keep down overtime expenses. The employee's response was to close the business by 9:00 p.m. whatever the cost. Now the corporate office is happy—they just don't realise they are losing customers and hundreds of dollars of future business. Much of the behaviour of employees towards their customers is the result of management philosophy.

The alternative management approach is to put the customer first and reward employees for serving the customer well. Roger Dow, Marriott's vice president of sales and marketing services, remarks: "We used to reward restaurant managers for things that were important to us, such as food costs. When have you heard a customer ask for the restaurant's food costs? You have to reward for what customers want from your business."

Conclusions

In this chapter, we have studied the relationships in the global hospitality and tourism industry, and we have also discussed why the marketing concept works well through a customer orientation. We also dilated on the concept of the lifetime value of a customer in relation to customer loyalty and retention.

Activity

1. Explain why marketing is important in the tourism and hospitality industry.
2. Explain why customer orientation is a key factor for the success of the tourism industry and the role of top management in achieving the same.
3. Briefly explain the concept of product management in the hospitality industry and how to overcome product-related difficulties in the industry.
4. Explain the concept of customer value and satisfaction and its relevance in the hospitality and tourism industry.

References

Achrol, R. S., & Kotler, P. (2022). Distributed marketing networks: The fourth industrial revolution. *Journal of Business Research, 150*, 515–527.

Dev, C. S., Buschman, J. D., & Bowen, J. T. (2010). Hospitality marketing: A retrospective analysis (1960–2010) and predictions (2010–2020). *Cornell Hospitality Quarterly, 51*(4), 459–469. https://doi.org/10.1177/1938965510376353

Hsu, M. J., Ting, H., Lui, T. W., Chen, S. C., & Cheah, J. H. (2022). Guest editorial: Challenges and prospects of AIoT application in hospitality and tourism marketing. *Journal of Hospitality and Tourism Technology, 13*(3), 349–355.

Ionel, M. (2016). Hospitality industry. *Ovidius University Annals: Economic Sciences Series, 1*(1), 187–191.

Jager, W. (2007). The four P's in social simulation, a perspective on how marketing could benefit from the use of social simulation. *Journal of Business Research, 60*(8), 868–875.

Langat, N. (2016). *Influence of Product, Price,Promotion and Place on Enterprise Project Performance: A Case of Safaricom Enterprise Project, Uasin Gishu County, Kenya* (Doctoral dissertation, University of Nairobi).

Oh, H., & Kim, K. (2017). Customer satisfaction, service quality, and customer value: Years 2000–2015. *International Journal of Contemporary Hospitality Management, 29*(1), 2–29.

Chapter 3

Service Characteristics of Hospitality and Tourism Marketing

Chapter Outline

- Introduction
- Service Culture
- Characteristics of Service Marketing

Chapter Objectives

After reading this chapter, you should be able to:

- Describe a service culture.
- Identify and explain the service characteristics that affect the marketing of a hospitality or tourism product.
- Explain the implications of the service characteristics for the marketing of hospitality and tourism services.

Introduction

Marketing initially developed in connection with selling physical products, such as toothpaste, cars, steel, and equipment. But today, one of the

DOI: 10.4324/9781032688497-3

major trends in many parts of the world is the phenomenal growth of services or products with little or no physical content. Services are different from tangible products in many ways and thus, need to be handled differently. Generally, a tangible product could be described as an object, or a thing as opposed to a service which is a deed, or a performance. This section, therefore, takes a look at the characteristics of hospitality and tourism services.

Service Culture

Some managers think of their operations only in terms of tangible goods. Thus, managers of fast-food restaurants who think they sell only hamburgers often have "slow, surly service personnel, dirty unattractive facilities, and few return customers." One of the most important tasks of a hospitality business is to develop the service side of the business, specifically, a strong service culture. Service culture is based on a customer-centric philosophy. Companies with a service culture prioritise the needs of the customer and tend to direct their mission, policies, systems, and processes towards satisfying the customer first.

A vibrant service culture empowers employees to solve customer problems. It is supported by a reward system based on customer satisfaction. Human beings generally do what is rewarded. If an organisation wants to deliver a quality product, the organisation's culture must support and reward customer need attention. A strong service culture also results in customer satisfaction, loyalty, and repeat visits which are essential for business success. The Azalaï Hotels Group, the first hotel chain in West Africa and a regional leader in the hotel sector, for instance, owes a lot of its success to its service culture. The hotel has developed a strategy geared towards satisfying the growing needs of its customers by constantly improving the quality of its high-end service through improved facilities and high-quality service guaranteed by African hospitality.

Characteristics of Service Marketing

Service marketers must be concerned with four characteristics of services: intangibility, inseparability, variability, and perishability.

Intangibility

Unlike physical products, services cannot be seen, tasted, felt, heard, or smelled before purchases are made. Prior to boarding an aeroplane, airline passengers have nothing but an airline ticket and the promise of safe delivery to their destinations. Members of a hotel sales force cannot take a hotel room with them on a sales call. In fact, they do not sell a room; instead, they sell the right to use a room for a specific period. When hotel guests leave, they have nothing to show for the purchase, but a receipt. Robert Lewis has observed that someone who purchases a service may go away empty-handed, but they do not go away empty-headed. They have memories that can be shared with others. Marriott Vacation Club International realises this, and they have made a deliberate effort to create memorable guest experiences. Marriott realises that a white-water rafting trip can create memories that a family visiting their Mountainside Resort in Utah will talk about for years. The fun the family had in experiencing white-water rafting, along with other experiences at the resort, will make them want to return. In the hospitality and tourism industry, many of the products sold are intangible experiences (Bebko, 2000).

If we are going to buy a car, we can take it for a test drive; if we are going to buy a meal at a restaurant, we do not know what we will receive until we have experienced the food and service. To reduce uncertainty caused by service intangibility, buyers look for tangible evidence that will provide information and confidence about the service. The exterior of a restaurant is the first thing that an arriving guest sees. The condition of the grounds and the overall cleanliness of the restaurant provide clues as to how well the restaurant is run. Tangibles provide signals as to the quality of the intangible service. The Regent Hotel in Hong Kong makes sure that all its uniformed and non-uniformed employees reinforce the hotel's image of elegance and professionalism. The appearance of the employees is part of the Regent's tangible evidence. This hotel also purposely parks luxury automobiles such as a Rolls-Royce in front to deliver an instant message of quality and upscale service (Moon, 2013).

In view of the intangible nature of hospitality and tourism products, marketers must thus provide *tangible clues* as to the quality of their products. This is also known as tangible evidence and includes the appearance of frontline staff, the landscaping around a hotel, the features and general design of a restaurant's website, the garnishing of a meal in a restaurant and the setup and decoration of a meeting room at a convention centre. Since customers cannot taste the service before purchase, tangibilising the service will provide some assurance of quality.

Inseparability

In most hospitality services, both the service provider and the customer must be present for the transaction to occur. Customer-contact employees are part of the product. The waitress in the restaurant is an inseparable part of the service offering. The food in a restaurant may be outstanding, but if the service person has a poor attitude or provides inattentive service, customers will down-rate the overall restaurant experience. They will not be satisfied with their experience. Service inseparability also means that customers are part of the product. A couple may have chosen a restaurant because it is quiet and romantic, but if a group of loud and boisterous conventioneers is seated in the same room, the couple will be disappointed. Managers must manage their customers so that they do not create dissatisfaction for other customers (Darmawan & Grenier, 2021).

Another implication of inseparability is that customers and employees must understand the service delivery system since they are both co-producing the service. Customers must understand the menu items in a restaurant so that they get the dish they expect. Hotel customers must know how to use the phone system and express checkout on the television. This means hospitality and travel organisations have to train customers just as they train employees. The Holiday Inn Newark is popular with international tourists who have just arrived from overseas. Many of these guests pay in cash or with travellers' cheques, as they do not use credit cards. On more than one occasion, the front-desk staff has been observed answering the phone of an upset guest who claims that their movie system does not work. The clerk must explain that they did not establish credit, because they paid for the room only and therefore have to come to the front desk and pay for the movie before it can be activated. Guests obviously become upset upon receiving this information. The hotel could avoid this problem and improve customer relations by asking guests at arrival time if they would like to deposit anything they might charge, such as in-room movies. It is also imperative for hospitality and tourism businesses to train their staff adequately for them to be able to deliver quality service. Sierra and McQuitty (2005) also posit that inseparability produces customer perceptions of shared responsibility for service outcomes, resulting in greater emotions.

Variability

Services are highly variable. Their quality depends on who provides them and when and where they are provided. There are several causes of service variability. Services are produced and consumed simultaneously, which limits quality control. Fluctuating demand makes it difficult to deliver consistent products during periods of peak demand. The high degree of contact between the service provider and the guest means that product consistency depends on the service provider's skills and performance at the time of the exchange. A guest can receive excellent service one day and mediocre service from the same person the next day. In the case of mediocre service, the service person may not have felt well or perhaps experienced an emotional problem. Lack of communication and heterogeneity of guest expectations is another source of variability. A restaurant customer ordering a medium steak may expect it to be cooked all the way through, whereas the person working on the broiler may define medium as having a warm, pink centre. The guest will be disappointed when he or she cuts into the steak and sees pink meat. Restaurants have solved this cause of variability by developing common definitions of steak doneness and communicating them to employees and customers. Sometimes the communication with the customer is verbal or sometimes printed on the menu. Though standardising services is a daunting task, some hospitality businesses, notably the fast-food restaurants have adopted a manufacturing approach which enables them to standardise their services. McDonald's for instance has been able to standardise its menu by employing the production-line approach. This involves the use of a restricted menu, division of labour, and a homogenous product like French Fries. The production processes are centralised. Pre-cut, partially cooked, and frozen potatoes are distributed to various outlets and staff are trained on how to deep-fry using standard equipment.

Customers usually return to a restaurant because they enjoyed their last experience. When the product they receive is different and does not meet their expectations on the next visit, they often do not return. Variability or lack of consistency in the product is a major cause of customer disappointment in the hospitality industry (Chiang et al., 2014).

Perishability

Services cannot be stored. A 100-room hotel that only sells 60 rooms on a particular night cannot inventory the 40 unsold rooms and then sell 140 rooms the next night. Revenue lost from not selling those 40 rooms is gone forever. Hotels can address the problem of perishability through reservations and creative pricing strategies. Because of service perishability, some hotels charge guests holding guaranteed reservations even when they fail to check into the hotel. Restaurants are also starting to charge a fee to customers who do not show up for a reservation. They, too, realise that if someone does not show up for a reservation, the opportunity to sell that seat may be lost. Another way of dealing with the risks associated with the perishability of hospitality and tourism services is by adopting a dynamic pricing strategy. This allows businesses to adjust the prices of their products in response to changing demand. Also, if services are to maximise revenue, they must manage capacity and demand because they cannot carry forward unsold inventory. The characteristic of perishability means that capacity and demand management are important to the success of a hospitality or travel company (Kefalas, 2019).

Summary

In this chapter, we have studied service culture and identified four service characteristics that affect the marketing of a hospitality or travel product.

Activity

1. What is service culture and what characteristics of services will you focus on building a service culture?
2. What is the relevance or importance of service culture in achieving success in the hospitality and tourism industry?
3. Assuming you are an executive in the hospitality industry what strategies you will put in place to deal with EACH of the unique characteristics of services in the industry.

References

Bebko, C. P. (2000). Service intangibility and its impact on consumer expectations of service quality. *Journal of Services Marketing, 14*(1), 9–26.

Chiang, F. F., Birtch, T. A., & Cai, Z. (2014). Front-line service employees' job satisfaction in the hospitality industry: The influence of job demand variability and the moderating roles of job content and job context factors. *Cornell Hospitality Quarterly, 55*(4), 398–407.

Darmawan, D., & Grenier, E. (2021). Competitive advantage and service marketing mix. *Journal of Social Science Studies (JOS3)*, *1*(2), 75–80.
Kefalas, S. (2019). Quality service in the hospitality industry: Achieving effective service processes and designs. In *Strategic innovative marketing and tourism: 7th ICSIMAT, Athenian Riviera, Greece, 2018* (pp. 39–47). Springer International Publishing.
Moon, Y. J. (2013). The tangibility and intangibility of e-service quality. *International Journal of Smart Home*, *7*(5), 91–102.
Sierra, J. J., & McQuitty, S. (2005). Service providers and customers: Social exchange theory and service loyalty. *Journal of Services Marketing*, *19*(6), 392–400.

Chapter 4

People Management Strategies for Service Businesses

Chapter Outline

- Introduction
- Managing Differentiation
- Managing Service Quality
- Service Marketing Management

Chapter Objectives

By the end of this chapter, you should be able to:

- Understand some of the strategies for service businesses.
- Understand the concept of differentiation in services and strategies to deal with it.
- Explain service quality and how to achieve service quality in the H&T sector.
- Understand how these strategies can be applied to manage services in the hospitality and tourism industry.

Introduction

Service marketers can do several things to increase service effectiveness in the face of intrinsic service characteristics. In this section, we are

DOI: 10.4324/9781032688497-4

going to study some of the strategies available to managers for service management.

Just like manufacturing businesses, good service firms position themselves strongly in chosen target markets through efficient marketing. However, because services differ from tangible products, they often require additional marketing approaches. In a product business, products are fairly standardised and can sit on shelves waiting for a customer. But in a service business, the customer and frontline service employee, interact to create the service. Thus, service providers must work to interact effectively with customers to create superior value during service encounters. Effective interaction, in turn, depends on the skills of frontline service employees, and the service production and support processes backing these employees (Riley, 2014).

Successful service companies focus their attention on both their employees and customers. They understand the service–profit chain which links service firm profits with employee and customer satisfaction. This chain consists of five links:

- Healthy service profits and growth—superior service firm performance.
- Satisfied and loyal customers—satisfied customers who remain loyal, repeal purchases, and refer other customers.
- Greater service value—more effective and efficient customer value creation and service delivery.
- Satisfied and productive service employees—more satisfied, loyal, and hardworking employees.
- Internal service quality—superior employee selection and training, a high-quality work environment, and strong support for those dealing with customers.

Therefore, reaching service profits and growth goals begins with taking care of those who take care of customers. JW Marriott, the founder of Marriott Hotels and Resorts was an ardent proponent of this notion. He believed in treating employees the same way managers want their customers to be treated. Nickson (2013) believes that people are the industry's most important asset. This popular mantra "if you take care of your associates they will take care of the guests" became a core ideology around which the Marriott hotel chain operates. Today, Marriott is one of the largest hotel chains in the world with 30 brands operating over 8,000 properties in 139 countries.

Taking care of employees is internal marketing which means that the service firm must effectively train and motivate its customer-contact employees and all the supporting service people to work as a team to provide customer satisfaction. For the firm to deliver consistently high-service quality, everyone must practice a customer orientation. It is not enough to have a marketing department doing traditional marketing while the rest of the company goes its way. Marketers also must get everyone else in the organisation to practice marketing. In fact, internal marketing must precede external marketing. We will discuss internal marketing in detail later in this module.

Interactive marketing means that perceived service quality depends heavily on the quality of the buyer–seller interaction during the service encounter. In product marketing, product quality often depends little on how the product is obtained. But in services of marketing, service quality depends on both the service deliverer and the quality of the delivery. The customer judges service quality not just on technical quality (the quality of the food) but also on its functional quality (the service provided in the restaurant). Service employees have to master interactive marketing skills or functions as well.

Today, as competition and costs increase, and as productivity and quality decrease, more marketing sophistication is needed. Service companies face the task of increasing three major marketing areas: their competitive differentiation, service quality, and productivity.

Managing Differentiation

In these days of intense price competition, service marketers often complain about the difficulty of differentiating their services from those of competitors. To the extent that customers view the services of different providers as similar, they tend to focus more on the price than the provider.

The solution to price competition is to develop a differentiated offer, delivery, and image. The offer can include innovative features that set one company's offer apart from its competitors'. For example, airlines have introduced such innovations as in-flight movies, advanced seating, air-to-ground telephone service, and frequent-flyer award programmes to differentiate their offers. British Airways even offers international travellers a sleeping compartment, hot showers, and cooked-to-order breakfasts. For organisations to succeed at product/service differentiation, however, the difference or uniqueness of the product must be of value to customers. For instance, Subway has capitalised on the rising demand for healthy diets by offering a healthier alternative to food provided by the traditional fast-food restaurants. Unfortunately, most service innovations are copied easily. Still, the service company that innovates regularly usually will gain a succession of temporary advantages and an innovative reputation that may help it keep customers who want to go with the best (Nickson, 2013; Kusluvan et al., 2010).

Managing Service Quality

One of the major ways that a service firm can differentiate itself is by delivering consistently higher quality than its competitors. Like manufacturers before them, many service industries have now joined the total quality movement. Many companies are finding that outstanding service quality can give them a potent competitive advantage that leads to superior sales and profit performance. Hospitality firms like Ritz-Carlton and Hilton have become almost legendary for their high-quality service. Lewis et al. (2016) are of the view that managing service quality is concerned with understanding what is meant by service quality, what its determinants are, and how they may be measured, and identifying the potential shortfalls in service quality and how they can be recovered.

The key is to exceed the customers' service-quality expectations. As the chief executive at American Express puts it, "Promise only what you can deliver and deliver more than you promise!" These expectations are based on past experiences, word of mouth, and service-firm advertising. If the perceived service of a given firm exceeds the expected service, customers are apt to use the provider again. Customer retention is perhaps the best measure of quality—a service firm's ability to hang onto its customers depends on how consistently it delivers value to them. Thus, whereas the manufacturer's quality goal might be zero defects, the service provider's goal is zero customer defections.

The service provider needs to identify the expectations of target customers concerning service quality. Unfortunately, service quality is harder to define and judge than product quality. Moreover, although greater service quality results in greater customer satisfaction, it also results in higher costs. Still, investments in service usually pay off through increased customer retention and sales. Whatever the level of service provided, it is important that the service provider clearly defines and communicates that level so that its employees know what they must deliver and customers know what they will get.

Many service companies have invested heavily to develop streamlined and efficient service-delivery systems. They want to ensure that customers will receive consistently high-quality service in even service encounters. Unlike product manufacturers who can adjust their machinery and inputs until everything is perfect, service quality will always vary, depending on the interactions between employees and customers. Problems inevitably will occur. As hard as they try, even the best companies will have an occasional late delivery, burned steak, or grumpy employee. However, although a company cannot always prevent service problems, it can learn to recover from them. Good service recovery can turn angry customers into loyal ones. In fact, good recovery can win more customer purchasing and loyalty than if things had gone well in the first place. Therefore, companies should take steps not only to provide good service every time but also to recover from service mistakes when they do occur.

The first step is to empower frontline service employees—to give them the authority, responsibility, and incentives they need to recognise, care about, and tend to customer needs; for example, Marriott has put some 70,000 employees through empowerment training which encourages them to go beyond their normal jobs to solve customer problems. Such empowered employees can act quickly and effectively to keep service problems from resulting in lost customers. Marriott Desert Springs revised the job description for its customer-contact employees. The major goal of these positions now is to ensure that "our guests experience excellent service and hospitality while staying at our resort." Well-trained employees are given the authority to do whatever it takes, on the spot, to keep guests happy. They are also expected to help management ferret out the cause of guests' problems, and to inform managers of ways to improve overall hotel service and guests' comfort (Nguyen et al., 2020)

Service Marketing Management

Service marketing management refers to the process of planning, organising, and implementing various marketing and innovative activities within a service organisation, to exceed customers' expectations. The primary purpose of a service organisation is to create customer value by anticipating and identifying customer requirements within the marketing environment, developing an appropriate service delivery system to satisfy and exceed these requirements, and ultimately making a profit. One core difference between managing the function of service and that of manufactured goods is that whereas a salesperson for example might not need in-depth knowledge of the manufacturing process to market the product, the reverse is the case for service. The marketer must be familiar with the service production process to be able to market it effectively. Generally, it is regarded as the perception of the service quality that is experienced by the tourists during their stay which is measured based on the services that they still remember after they have returned to their hometown (Abdulla et al., 2020).

Guaranteeing customer satisfaction in this scenario is a major challenge. In the case of the service industry, the customer first needs to develop trust in the service organisation before he/she buys the service. The client may place more importance on the amount of faith in the service organisation, than the services being offered and their value proposition. Most product companies have dedicated sales staff, while in the service industry, the service providers usually engage in the simultaneous production and delivery of services. Coordinating marketing, operations, and human resource efforts can be a challenging task in the delivery of service.

Summary

In this chapter, we looked at some of the strategies for service businesses and we also pointed out how these strategies can be applied to manage services in the hospitality and tourism industry.

Activity

1. State and explain key service marketing strategies in the H&T sector and explain how these strategies help improve service quality in the tourism and hospitality industry in Africa.
2. Define differentiation in service marketing and provide some strategies managers in the H&T industry could put in place to deal with service differentiation.
3. Explain service quality and strategies managers can employ to achieve service quality in the H&T industry.
4. Explain the concept of people management in the service industry and its relevance in achieving organisational success, particularly in the H&T industry.

References

Abdulla, S. A. M., Khalifa, G. S., Abuelhassan, A. E., Nordin, B. B., Ghosh, A., & Bhaumik, A. (2020). Advancement of destination service quality management technology in tourism industry. *Journal of Critical Reviews*, *7*(11), 2317–2324.

Kusluvan, S., Kusluvan, Z., Ilhan, I., & Buyruk, L. (2010). The human dimension: A review of human resources management issues in the tourism and hospitality industry. *Cornell Hospitality Quarterly*, *51*(2), 171–214.

Nguyen, H. T. T., Nguyen, N., & Pervan, S. (2020). Development and validation of a scale measuring hotel website service quality (HWebSQ). *Tourism Management Perspectives*, *35*, 100697.

Nickson, D. (2013). *Human resource management for hospitality, tourism and events*. Routledge.

Riley, M. (2014). *Human resource management in the hospitality and tourism industry*. Routledge.

Chapter 5

The Role of Marketing in Strategic Planning

Chapter Outline

- Market-oriented Strategic Planning
- Corporate Strategic Planning
- Defining the Corporate Mission
- Establishing Strategic Business Units
- Developing Growth Strategies
- Intensive Growth
- Diversification Growth

Chapter Objectives

At the end of this chapter, you should be able to:

- Explain company-wide strategic planning.
- Explain the concepts of stakeholders, processes, resources, and organisations as they relate to high performance.
- Explain the four planning activities of corporate strategic planning.
- Describe the processes involved in defining a company mission and setting goals and objectives.

DOI: 10.4324/9781032688497-5

Market-oriented Strategic Planning

Market-oriented strategic planning is the managerial process of developing and maintaining a feasible fit between the organisation's objectives, skills, and resources and its changing market opportunities. Strategic planning aims to help a company select and organise its businesses in a way that keeps the company healthy despite unexpected upsets occurring in any of its specific businesses or product lines. When managers develop a good marketing plan or business plan, the principles of strategic planning must be clearly understood (Taghian, 2010).

Three key ideas define strategic planning. The first calls for managing a company's businesses as an investment portfolio, for which it will be decided whether the business entities deserve to be built, maintained, phased down (harvested, milked), or terminated. The second key idea is to assess accurately the future profit potential of each business by considering the market's growth rate and the company's position and fit. It is not sufficient to use current sales or profits as a guide. For example, if Hyatt, Marriott, and Holiday Inn had used only current profits as a guide to investment opportunities, they would have continued to invest primarily or solely in commercial hotels in downtown and airport locations and, in the case of Holiday Inns, solely in family motels. Instead, Marriott offers a diverse portfolio of brand-name lodgings, such as Marriott Marquis, Courtyard, Fairfield Inn and Suites, and Residence Inn, for different market segments. Hyatt has been very active in resort development and is a recognised leader in programmes for children and teenagers with Camp Hyatt and Rock Hyatt. Holiday Inn has a diversified portfolio of hospitality products, including casino hotels and Holiday Inn Express.

The third key idea underlying strategic planning is that of strategy. For each business, the company must develop a game plan for achieving its long-run objectives. Furthermore, no one strategy is optimal for all competitors in that business. Each company must determine what makes the most sense in light of its industry position and its objectives, opportunities, skills, and resources. Thus, in the airline industry, American Airlines is pressing for cost reduction as a full-service airline and a strong global market share. Southwest continues to strive for low-cost, limited domestic service while acquiring other carriers, such as Mark Air, with similar strategies. The future of the airline industry remains uncertain, but the strategies of both carriers could prove to be correct. Marketing and strategic planning should be viewed as partnerships contributing to the long-run success of a hospitality firm.

To understand strategic planning, we need to recognise that most large companies consist of four organisational levels: corporate, division, business, and product. In recent years, hospitality companies have taken on the organisational appearance of multi-levels. Corporate headquarters is responsible for designing a corporate strategic plan to guide the whole enterprise into a profitable future; it makes decisions on how much resource support is to be allocated to each division, as well as which businesses to start or eliminate. Each division establishes a division plan covering the allocation of funds to each business unit within that division. Each business unit in turn develops a business unit strategic plan to carry that business unit into a profitable future. Finally, each product level (product line, brand) within a business unit develops a marketing plan for achieving its objectives in its product market. These plans are then implemented at various levels of the organisation, results are monitored and evaluated, and corrective actions are taken (Fernandes et al., 2020).

Stakeholders

The starting point for any business is to define the stakeholders and their needs. Traditionally, most businesses primarily nourish their stockholders. Today's businesses increasingly recognise that unless other stakeholders, customers, employees, suppliers, and distributors are nourished, the business may never earn sufficient profits for the stockholders. Maintaining good relationships with stakeholders is seen as a main source of competitive advantage which leads to trust, goodwill, reduced uncertainty, improved business dealings, and ultimately higher firm performance. Thus, if British Airways employees, customers, dealers, and suppliers are unhappy, profits will not be achieved. This leads to the principle that a business must at least strive to satisfy the minimum expectations of each stakeholder group. To ensure accountability and mitigate harm, it is critical that diverse stakeholders can interrogate black-box automated systems and find information that is understandable, relevant, and useful to them (Suresh et al., 2021).

It has been suggested that an employee stakeholder group, which may be of increased importance to hospitality firms, is women with small children who do not wish to work full-time. Many do not want to leave the workplace entirely but are forced to do so as their employers do not permit flexible scheduling and reduced work hours. Many of these individuals could provide valuable contributions but find it impossible to do so under the current organisational structure and policy.

Processes

Company work is traditionally carried on by departments. However, departmental organisations pose some problems. Departments typically operate to maximise their objectives, not necessarily the company's. Walls come up between departments, and there is usually less-than-ideal cooperation. Work is slowed and plans often are altered as they pass through departments. Business Process Management (BPM) can be perceived as a set of methods, techniques, and tools used for identifying, analysing, optimising (redesigning), and monitoring processes with a focus on productivity increase and cost reduction (Dumas et al., 2013; Chalupa & Petricek, 2020).

Companies are increasingly refocusing their attention on the need to manage processes even more than departments. This is even more imperative in hospitality and tourism establishments where the delivery of quality service requires the collaboration of different personnel and departments. In a typical hotel, the various departments including the front office, marketing and sales, housekeeping, food and beverage, maintenance, security, and human resources must work together. Housekeeping and the front office communicate with each other about the status of rooms. Also, the front office furnishes the marketing and sales department with data on guest histories, details concerning each guest's visit, etc. In view of this, hospitality and tourism firms are studying how tasks pass from department to department and the impediments to effective output. They are now building cross-functional teams that manage core business processes.

The Las Vegas Hilton was concerned with the profit contribution from various market segments and how to deal with this issue. The result was a radically different approach to hotel accounting called market segment accounting. This new approach incorporated marketing and strategic planning into accounting rather than viewing them as separate, stand-alone areas, and philosophies.

Resources

To carry out processes, a company needs such resources as personnel, materials, machines, and information. Traditionally, companies sought to own and control most of the resources that entered the business. Now that is changing. Companies are finding that some resources under their control are not performing and those they could outsource. More companies today have decided to outsource less critical resources. On the other hand, they appreciate the need to own and nurture those core resources and competencies that make up the essence of their business. Smart companies are identifying their core competencies and using them as the basis for their strategic planning. The most critical resource of hospitality and tourism businesses is the human resource. It is not enough for such businesses to recruit people with the requisite attitude, experience, and skills. They can also nurture such skills through training and development.

Organisation

The organisational aspect of a company consists of its structure, policies, and culture, all of which tend to become dysfunctional in a rapidly changing company. Although structure and policies can be changed, the company's culture is the hardest to change. Companies must work hard to align their organisational structure, policies, and culture to the changing requirements of business strategy.

Corporate Strategic Planning

Corporate headquarters has the responsibility for wetting into motion the whole planning process. Some corporations give a lot of freedom to their business units but let them develop their strategies; others set the goals and get heavily involved in the individual strategies. The hospitality industry faces the need for greater empowerment of employees, particularly at middle-management levels due to the simultaneity of production and consumption as well as the locational peculiarities of different business units. For instance, you don't expect the Steeers restaurant in Cape Town to operate in the same way as the one in Nairobi, as the products and services have to be adapted to the needs and characteristics of the target market. When McDonald's entered the Indian market for instance, it had to replace its famed beef-based Big Mac with a mutton-based Maharajah Mac due to the fact that about 80 percent of the population revere cows. It has been suggested that many of the traditions within the hospitality industry have experienced little change (Laury et al., 2020).

Next, we examine four planning activities that all corporate headquarters must undertake:

1. Defining the corporate mission.
2. Establishing strategic business units (SBUs).
3. Assigning resources to each SBU.
4. Planning new businesses.

Defining the Corporate Mission

A hospitality organisation exists to accomplish something: to provide warmth, to provide accommodation, to provide food and drinks, to entertain, and so on. Its specific mission or purpose is usually clear at the beginning. Over time, some managers may lose interest in the mission or the mission may lose its relevance in light of changed market conditions.

When the management senses that the organisation is drifting, it must renew its search for purpose. According to Peter Ducker an Austrian-born American management consultant, it is time to ask some fundamental questions. What is our business? Who is the customer? What is value to the customer? What will our business be? What should our business be? These simple-sounding questions are among the most difficult the company will ever have to answer. Successful companies raise these questions continuously and answer them thoughtfully and thoroughly Drucker (2012)

The company's mission is shaped by history. Marriott is a classic example of a business whose mission has been strongly informed by the vision of its founding fathers. Marriott's mission, of enhancing the lives of its customers by creating and enabling unsurpassed vacation and leisure experience is based on the founding father's vision of building the world's favourite travel company and of leaving a legacy of excellence in the service industry. Every company has a history of aims, policies, and achievements. The organisation must not depart too radically from its history. The organisation's resources determine which missions are possible. Singapore Airlines would be deluding itself if it adopted the mission to become the world's largest airline.

Finally, the organisation should base its mission on its distinctive competencies. McDonald's could probably enter the solar energy business, but that would not use its core competence, that is, providing low-cost food and fast service to large groups of customers.

Organisations develop corporate mission statements to share with their managers, employees, and, in many cases, customers and other public. A well-worked-out mission statement provides company employees with a shared sense of purpose, direction, and opportunity. Writing a formal mission statement is not easy. Some organisations spend a year or two trying to prepare a satisfactory statement about their company's purpose. Good mission statements embody many characteristics. They should focus on a limited number of goals. The mission statement should define the major competitive scopes within which the company will operate (Drucker & Maciariello, 2008).

Establishing Strategic Business Units

Some corporate entities span operate several businesses. However, they sometimes fail to define them carefully. Businesses are too often defined in terms of products. Hospitality and tourism companies could be in the "hotel business'," "foodservice business," or the "cruise line business." However, market definitions of a business are superior to product definitions. A business must be viewed as a customer-satisfying process. Not a goods-producing process. Hospitality and tourism companies should define their business in terms of customer needs, not products. Ultimately**,** the most important purpose of any product or service is to satisfy the customer's needs and not the organisation's needs (Khaustova et al., 2020).

The BCG matrix also called the growth share matrix is a portfolio management framework that helps companies decide how to prioritise their different businesses. It is a table, split into four quadrants, each with its own unique symbol that represents a certain degree of profitability: question marks, stars, pets (often represented by a dog), and cash cows. By assigning each business to one of these four categories, executives could then decide where to focus their resources and capital to generate the most value, as well as where to cut their losses.

Having plotted its various businesses in the growth-share matrix, the company then determines whether its business portfolio is healthy. An unbalanced portfolio would have too many dogs or question marks and/or too few stars and cash cows. The company's next task is to determine what objective, strategy, and budget to assign to each SBU. Four alternative objectives can be pursued.

Build: The objective is to increase the Sub's market share, even foregoing short-term earnings to achieve this. Building is appropriate for question marks whose shares have to grow to become stars.

Hold: In this instance, the objective is to preserve the Sub's market share. This objective is appropriate for strong cash cows if they are to continue to yield a large positive cash flow.

Harvest: This looks at increasing the Sub's short-term cash flow regardless of the long-term effect. This strategy is appropriate for weak cash cows whose future is dim and from whom more cash flow is needed. Harvesting can also be used with question marks and clogs.

Divest: This seeks to sell or liquidate the business because resources can be better used elsewhere.

As time passes, Slues change their position in the growth-share matrix. Successful SBUs have a life cycle. They start as question marks, became stars, then cash cows, and, finally, dogs. For this reason, companies should examine not only the current positions of their businesses in the growth-share matrix but also their moving positions. Although the portfolio is basically healthy, wrong objectives or strategies could be assigned. The worst mistake would be to require all the SBUs to aim for the same growth rate or return level. The very point of SBU analysis is that each business has a different potential and requires its own objective. Additional mistakes would include the following:

1. Leaving cash-cow businesses with too little in retained funds, in which case they grow weak, or leaving them with too much in retained funds, in which case the company fails to invest enough in new growth businesses.
2. Making major investments in dogs hoping to turn them around, but failing each time.
3. Maintaining too many question marks and underinvesting in each. Question marks should either receive enough support to achieve segment dominance or be dropped.

Just as a corporation uses the BCG grid to manage its businesses, a restaurant manager can apply the concepts of the BCG grids to manage individual menu items. An application of the BCG matrix to the menu of McDonald's showed that McChicken, Fish-o-Fillet burgers, and Fries from which the company makes lots of money are its cash cows. Ice Cream and Cone were found to be Question Marks and Dogs because though it has the lowest market share, it has the highest market growth rate. However, if the company invests too much money in this product, it would be better to convert it into a star else it has to be liquidated as it would negatively impact the whole product's life cycle. McFlurry which has a high market share and maximum market growth rate was found to be the company's star product as the company makes a profit from it. Katakana and Smith used a similar matrix for menu planning. Their menu-engineering matrix has plow horses: high-volume, low-contribution items; stars: high-volume, high-contribution items; puzzles: high-contribution, low-volume items; and dogs: low-volume, low-contribution items. The models used for strategic analysis can often be adapted and used at the business, division, and product levels (Kader & Hossain, 2020; Clarissia, 2020; Tran, 2020).

Developing Growth Strategies

Marketing has the main responsibility for achieving profitable growth for the company. Marketing must identify, evaluate, and select market opportunities and lay down strategies for capturing them. Ansoff has proposed a useful framework for detecting new intensive growth opportunities. Called the Ansoff product-market expansion grid, the management first considers whether it could gain more market share with its current products in their current markets (market penetration strategy). Next, it considers whether it can find or develop new markets for its current products (market development strategy). Then it considers whether it can develop new products of potential interest to its current markets (product development strategy). We apply it here to Starbucks.

Intensive Growth

Firms that pursue intensive growth strategies expand their product lines or market reach. Intensive growth can be best explained with a reference to the Ansoff product market expansion grid. There are three intensive growth options, namely, market penetration strategy, market development strategy, and product development strategy.

Market Penetration: First, Starbucks management might consider whether the company can achieve deeper market penetration—making more sales to current customers without changing its products. It might add new stores in current market areas to make it easier for more customers to visit. Improvements in advertising, prices, service, menu selection, or store design might encourage customers to stop by more often or to buy more during each visit. For example, Starbucks recently began adapting its menu to local tastes around the country. Though Starbucks already operates in about 80 countries and territories, through the market penetration strategy, the company is opening more company-owned stores or licensed/franchised café locations. The company further seeks to open more stores in countries where it has a weak presence including Africa and the Middle East.

Market Development Strategy: Starbucks management might consider possibilities for market development—identifying and developing new markets for its current products. For instance, managers could review new demographic markets—such as senior consumers or ethnic groups—to see if new groups could be encouraged to visit Starbucks coffee shops for the first time or to buy more from them. Managers also could review new geographic markets. Starbucks is now expanding swiftly into new US markets, especially in the southeast and southwest. It is also developing its international markets, with stores popping up rapidly in Asia, Europe, and Australia.

Product Development Strategy: Next, management should consider product development—offering modified or new products to current markets. Product development strategy borders on innovation. Through product innovation, Starbucks now offers brewing equipment and ready-to-drink products at grocery stores. By examining these three intensive growth strategies, management will hopefully discover several ways to grow. Still, that may not be enough, in which case management must also examine diversification and integrative growth opportunities.

Diversification Growth

Diversification growth makes sense when good opportunities can be found outside the present businesses. A good opportunity is one where the industry is highly attractive and the company has the mix of business strengths to be successful. Three types of diversification can be considered. First, the company could seek new products that have technological and/or marketing synergies with existing product lines, even though the products may appeal to a new class of customers (concentric diversification strategy). This is a common strategy adopted by global hospitality firms like Accor, Marriott, Yum Brands, and McDonald's. MacDonald's has had a history of diversifying its menu to suit different markets. Through this strategy, it has created different menus like Pizza McPuff, McVeggie, McCafe, and McFlurry. Second, the company might search for new products that could appeal to its current customers, although technologically unrelated to its current product line (horizontal diversification strategy). Hotels, restaurants, cruise lines, and airlines all pursue this strategy when they sell gift items such as T-shirts, perfume, and luggage.

Summary

In this chapter, we discussed what company-wide strategic planning is, the concepts of stakeholders, processes, resources, and organisation as they relate to high-performing businesses. We also examined the processes involved in defining a company mission and setting goals and objectives.

Activity

1. Define what strategic planning is and outline the processes involved in defining a company mission.
2. How important is corporate planning to managers in the hospitality industry in relation to their mission?
3. Explain market-oriented strategic planning and its role in achieving organisational goals in the H&T industry.
4. Identify the common mistakes managers are likely to make when establishing strategic business units and state how those mistakes could be avoided.
5. Explain the various strategies managers in the H&T industry could implement to achieve growth of their organisations.
6. Briefly explain the four key planning activities corporate headquarters should embark on to achieve success in strategic planning.

References

Chalupa, S., & Petricek, M. (2020). The application of business process management in the hospitality industry: A case study. *IBIMA Business Review*, *1*(301930), 1–11.

Drucker, P. F. (2012). *Management challenges for the 21st century*. Burlington, MA: Butterworth-Heinemann.

Drucker, P. F., & Maciariello, J. A. (2008). *Management: Revised edition.* New York: Collins.

Dumas, M., Rosa, M. I., Mendling, J., & Reijers, H. A. (2013). *Fundamentals of business process management.* Berlin: Springer.

Fernandes Sampaio, C. A., Hernández Mogollón, J. M., & de Ascensão Gouveia Rodrigues, R. J. (2020). The relationship between market orientation, customer loyalty and business performance: A sample from the Western Europe hotel industry. *Tourism and Hospitality Research, 20*(2), 131–143.

Kader, M. A., & Hossain, H. (2020). An analysis on BCG growth sharing matrix. *International Journal of Economics, Business and Accounting Research (IJEBAR), 4*(1), 112–135.

Khaustova, Y., Durmanov, A., Dubinina, M., Yurchenko, O., & Cherkesova, E. (2020). Quality of strategic business management in the aspect of growing the role of intellectual capital. *Academy of Strategic Management Journal, 19*(5), 1–7.

Laury, H. A., Matondang, N., & Sembiring, M. T. (2020, May). Balanced scorecard in the integration of corporate strategic planning and performance: A literature review. In *IOP conference series: Materials science and engineering* (Vol. 801, No. 1, p. 012135). IOP Publishing.

Suresh, H., Gomez, S. R., Nam, K. K., & Satyanarayan, A. (2021, May). Beyond expertise and roles: A framework to characterize the stakeholders of interpretable machine learning and their needs. In *Proceedings of the 2021 CHI conference on human factors in computing systems* (pp. 1–16).

Taghian, M. (2010). Marketing planning: Operationalising the market orientation strategy. *Journal of Marketing Management, 26*(9–10), 825–841.

Chapter 6

Business Strategy Planning

Chapter Outline

- Introduction
- The Microenvironment of Hospitality and Tourism Businesses
- The Macroenvironment of Hospitality and Tourism Businesses
- Responding to the Marketing Environment
- Environmental Scanning
- Using Information about the Marketing Environment
- The Business Unit Strategic Planning Process
- SWOT Analysis
- Goal Formulation
- Strategy Formulation
- Programme Formulation
- Implementation
- Feedback and Control

Chapter Objectives

At the end of this chapter, you should be able to:

- List and discuss the importance of the elements of the company's microenvironment.
- Describe the macroevironmental forces that affect the company's ability to serve its customers.

DOI: 10.4324/9781032688497-6

- Explain how changes in the demographic and economic environments affect marketing, and describe the levels of competition.
- Identify the major trends in the firm's natural and technological environments.
- Explain the key changes that occur in the political and cultural environments.
- Discuss how companies can be proactive rather than reactive when responding to environmental trends.
- Discuss how to design business portfolios and growth strategies.
- Explain the steps involved in the business strategy planning process.

Introduction

Every business operates in an environment. This environment consists of forces that affect the operation of the business. The marketing environment of a hospitality business is made up of actors and forces within or outside the business organisation that affect its ability to develop and maintain successful transactions with its target customers. The marketing environment is composed of a microenvironment and a macroenvironment. The microenvironment consists of actors and forces within the company's immediate environment that can affect its ability to serve its customers, suppliers, marketing channel firms, customer markets, and a broad range of publics. The company has complete or some level of control over these actors. The macroenvironment, on the other hand, consists of the larger societal forces, outside the control of the company, that affect the entire microenvironment, that is, demographic, economic, natural, technological, political, and cultural forces. We first examine the company's microenvironment and then its macroenvironment.

The Microenvironment of Hospitality and Tourism Businesses

The task of marketing managers is to create attractive offers for target markets. To be able to do this, they need to understand the forces within the microenvironment. Various actors in its microenvironment will affect the degree of success of the company. For instance, the availability of qualified and well-motivated staff is crucial to the success of a hospitality firm. The actors in the microenvironment include the company, suppliers, market intermediaries, customers, and other publics (Hassanien et al., 2010).

The Company

Marketing managers do not operate in a vacuum. They must work closely with top management and the various departments in a company. The finance department is concerned with finding and using the funds required to carry out the marketing plan. The accounting department has to measure revenues and costs to help marketing know how well it is achieving its objectives. Product development is responsible for creating new products to fit the needs of the changing marketplace. Housekeeping is responsible for delivering clean rooms sold by the sales department. All company departments will have some impact on the success of the marketing plan.

Suppliers

Suppliers are firms and individuals that provide the resources needed by the company to produce its goods and services. These include suppliers of food, linen, utility services, and ICT service providers. Trends and developments affecting suppliers can, in turn, affect a company's marketing plan. Suppose that a restaurant manager decided to run a live lobster special for the weekend. The seafood supplier was called, who promised to supply 200 lobsters for the weekend promotion. However, on Friday morning, the supplier called to inform them that the cargo from Boston was short on lobsters and that they would not be delivered until Saturday morning. The manager must now find an alternative source of lobsters or disappoint guests who had made their reservations for Friday night.

In another case, Steak and Ale wanted to expand its menu with a new scallop seafood dish. It took the corporate office six months to perfect the scallop dish. During the development period, the price of scallops doubled. Due to the price hike, the restaurant would have charged a price higher than what customers would pay. The project was therefore abandoned. Marketing management must therefore pay attention to changes in supply availability (as affected by shortages and strikes) and supply costs.

Marketing Intermediaries

Marketing intermediaries are specialised groups of suppliers who help the company in promoting, selling, and distributing its goods to end users. They are business firms that help hospitality companies find customers or make sales. They include travel agents, tour operators, and hotel representatives.

For example, a tour operator creates travel packages that include air transportation, ground transportation, and hotel accommodations. These packages are promoted through various mediums such as newspaper advertising and travel agents. Through bulk purchasing, the tour operator receives reduced prices, which enables him/her to pay the travel agent a commission for selling the product, offer the customer a good price and produce a profit. In choosing wholesalers, hotels must select reputable firms that will deliver the promised product to the customer and pay the hotel for their services (Kracht & Wang, 2010).

In manufacturing industries, transportation systems help move products from the factory to the customers. However, in the case of the tourist product, it is rather the customer who goes for the product. Thus travel intermediaries like travel agents have to be located at places which are convenient to the customer. Also, unlike manufactured goods, the travel intermediary cannot hold stock of the tourist product. This is a result of the intangibility of the product. Closer collaboration between tourism service providers and intermediaries is required for successful marketing. It is also important to point out that due to advances in ICT, the industry is increasingly confronted with the issue of disintermediation. Through disintermediation, the services of intermediaries are dispensed with and consumers buy directly from producers. Many travellers are now able to plan their own trips by buying different aspects of the product such as accommodation, transport, and food directly from suppliers.

Marketing services agencies are suppliers that help the firm formulate and implement its marketing strategy and tactics by working closely with the company's marketing plan. These suppliers include public relations agencies, advertising agencies, and direct mail houses. They also include marketing research firms, media firms, and marketing consulting firms which help companies target and promote their products to the right target markets. These firms can vary in creativity, quality, service, and price. Because these marketing agencies are vital to the achievement of an organisation's marketing objectives, a hospitality company should regularly review the performance of its agencies and replace those that underperform.

Financial intermediaries include banks, creditors, insurance companies, and other firms that help hospitality companies finance their businesses or insure the risks associated with their businesses. Because rising credit costs, limited credit, or both can seriously affect a company's marketing performance, the company has to develop strong relationships with important financial institutions. However, the companies must not succumb to the unmanageable growth expectations of their financial intermediaries (Banna et al., 2021).

The Macroenvironment of Hospitality and Tourism Businesses

A hospitality company and its suppliers, marketing intermediaries, customers, and other publics operate in a larger microenvironment that presents opportunities and poses threats. The company must monitor and respond to these uncontrollable forces. Every company faces a broad range of competitors. According to the marketing concept, to be successful, a company must satisfy the needs and wants of consumers better than its competitors. Marketers must do more than simply adapt to the needs of target customers. They must also adapt to the strategies of their competitors who serve the same target markets. Companies must gain strategic advantage by strongly positioning their products in the minds of consumers.

No single competitive marketing strategy is best for all companies. Each firm must consider its size and industry position in relation to those of its competitors. For instance, the strategies adopted by large firms with dominant positions in an industry may not be the same as those of smaller firms. For example, a large restaurant chain can capitalise on its buying power to purchase national advertising, spreading the cost among hundreds or thousands of operations. But a small, independent restaurant can adjust quickly to local trends and can offer more menu variety because it does not have to standardise menu items across thousands of restaurants. Both large and small firms must find marketing strategies that give them specific advantages over competitors operating in their markets (Sijabat, 2019).

Demographic Environment

Demography is the study of human populations in terms of size, density, location, age, sex, race, occupation, and other statistics. The demographic environment is of significant interest to marketers in the hospitality industry because it is a people-oriented industry. Therefore, changes in the demographic environment have major implications for the industry. For instance, if the majority of the population is in their middle age and working as middle-class people, the possibility of these age group going for holidays, visiting tourist sites, and eating outside of the home is high since they have purchasing power and this will have a positive effect on the hospitality industry. The reverse is also true.

The Economic Environment

The economic environment is made up of economic factors that determine or affect the ability of customers to purchase products or services of an organisation. If the economic conditions such as income levels, and employment are positive, it increases the purchasing power and spending behaviour of customers. Companies use economic forecasting to monitor changes in major economic variables such as income, cost of living, interest rates, and inflation, which have a major impact on the marketplace.

The hospitality and tourism industry is highly susceptible to various macroeconomic factors, such as GDP, interest rates, and inflation. GDP growth indicates the overall health and strength of an economy, while interest rates make borrowing more affordable for individuals and businesses. Inflation which is the rate at which prices for goods and services rise, significantly impacts the industry. Moderate inflation is an indication of a growing economy and increased consumer purchasing power, while high inflation erodes consumer buying power resulting in rising costs for businesses. Exchange rates can determine the affordability and attractiveness of destinations for international travellers, while government policies and regulations can either stimulate or hinder industry growth. Consumer spending patterns have a direct influence on the hospitality and tourism industry, and businesses must adapt to changing consumer behaviours and preferences to remain competitive.

The Technological Environment

The technology environment is probably the most dramatic force shaping our future. Technology has given us numerous benefits as vehicles, the television, and credit cards. It has also released such atrocities as nuclear missiles and assault guns. Technological advancements are one of the forces driving change in the the hospitality industry. The invention of the jumbo jet has made travel cheaper, faster, and easier. Computer Reservation Systems have enabled hotels to sell their capacity with relative ease. Mobile technology is changing the face of marketing and recent developments in Robotics and Artificial Intelligence are changing the customer experience.

New technologies such as radio-frequency identification (RFID), GPS, and Bluetooth are being used to track products and customers at various points in the distribution channel. Walmart has encouraged suppliers shipping products to its distribution centres to apply RFID tags to their pallets, and retailers such as American Apparel, Macy's, and Bloomingdales are now installing item-level radio frequency identification (RFID) systems in their stores. Many hotels are now using RFID for access control, asset tracking, payments, and inventory control. These technologies can provide many advantages to both buyers and sellers, but they can also be a bit scary.

Technological advancements have transformed consumer behaviour and expectations in the hospitality and tourism industry, and businesses need to embrace technology and adapt their operations to cater to digitally savvy customers. Offering online booking platforms, personalised digital experiences, and leveraging social media engagement can enhance customer satisfaction and convenience. Sociocultural factors are macroeconomic influences that significantly impact the ability of firms in the hospitality and tourism industry to satisfy their customers. Shifts in consumer preferences, such as the growing interest in experiential travel, adventure tourism, or wellness retreats, impact the types of products and services that customers seek.

Political Environment

The political environment plays a crucial role in shaping the operational landscape of the hospitality and tourism industry. Political factors include government policies, regulations, political stability, and geopolitical relationships. Government policies and regulations on visa acquisition, immigration, taxation, labour laws, health and safety regulations, and environmental sustainability can have significant implications for businesses. Stable political environments instil confidence in both domestic and international tourists, encouraging them to visit destinations. Geopolitical relationships and international relations such as bilateral agreements, trade policies, diplomatic relations, and political tensions between countries play a significant role in shaping the industry.

The political environment is a macroeconomic factor that significantly influences the ability of firms in the hospitality and tourism industry to satisfy their customers. Positive diplomatic relations and open borders can facilitate travel, while strained relationships or geopolitical conflicts can restrict tourism and affect customer satisfaction. Political cooperation and collaboration are essential for promoting seamless travel and fostering positive customer experiences. Crisis management and disaster preparedness are crucial in the hospitality and tourism industry, while sustainable development and environmental policies are becoming increasingly important. Understanding and adapting to the political environment is essential for industry stakeholders to navigate challenges, capitalise on opportunities, and meet customer expectations effectively. Collaboration between governments and industry players is vital in creating an enabling political environment that promotes customer satisfaction, enhances destination attractiveness, and drives sustainable growth.

The Sociocultural Environment

Sociocultural factors including demographics, lifestyles, values, attitudes, and behaviours play a significant role in shaping the hospitality and tourism industry. For companies in the hospitality and tourism industry to effectively satisfy their customers, it is imperative for them to understand and adapt to these factors. Demographic characteristics, such as age, gender, income levels, and family structures, greatly influence customer preferences and expectations. Cultural diversity and preferences are fundamental aspects of the industry, and businesses need to offer a variety of products and services that cater to different cultural backgrounds. Consumer lifestyles and values influence their choices and expectations when it comes to hospitality and tourism experiences.

Tourism and hospitality businesses must understand and adapt to these sociocultural factors, tailor their service offerings to these factors, create personalised experiences, and provide customer-centric services. To achieve this, the businesses must first embrace diversity. Sociocultural factors significantly influence the ability of firms in the hospitality and tourism industry to satisfy their customers. Cultural diversity, changing demographics, consumer behaviours, social media, health and wellness trends, authenticity, and environmental responsibility all shape customer expectations and preferences. Firms that understand and adapt to these sociocultural dynamics can effectively cater to customer needs, deliver exceptional experiences, and build strong customer relationships.

Responding to the Marketing Environment

Many hospitality and tourism firms view the marketing environment as an "uncontrollable" element to which they must adapt. They passively accept the marketing environment and do not try to change it. They analyse environmental forces and design strategies that will help the company avoid threats and take advantage of the opportunities that the environment provides.

Other companies take an environmental management perspective by being proactive. Rather than simply watching and reacting to what happens in the marketing environment, these firms take aggressive action to affect the publics and the forces in their marketing environment. Some businesses hire lobbyists to influence legislation affecting their industries and stage media events to gain favourable press coverage. They run advertorials (ads expressing editorial points of view) to shape public opinion. They press lawsuits and file complaints with regulators to keep competitors in line. They form contractual agreements to control their distribution channels better.

One of the elements of the macroenvironment that can be influenced is the political environment. Large companies hire lobbyists to present their interests at the local, state, and federal levels of government. These companies, along with smaller companies, join trade organisations such as the American Hotel and Lodging Association (AH&LA), Ghana Hotels Association (GHA), and the Kenya Tourism Federation (KTF). The trade associations also hire lobbyists and form Political Action Committees (PACs) to represent and communicate their industry's concerns to government officials. By communicating the possible effects of proposed legislation on the industry and the community, PACs can sometimes influence pending legislation.

Marketing management cannot always affect environmental forces; in many cases, it must settle for simply watching and reacting to the environment. For example, a company would have little success trying to influence geographic population shifts, the economic environment, or major cultural values. But whenever possible, smart marketing managers take a proactive rather than a reactive approach to the publics and forces in their marketing environment.

Environmental Scanning

Environmental scanning which involves gathering data on events and trends in an organisation's internal and external environment has proved beneficial to many hospitality companies. It follows the following steps:

(1) Determine the environmental areas that need to be monitored.

(2) Determine how the information will be collected, including information sources, the information frequency, and who will be responsible.

(3) Implement the data collection plan.

(4) Analyse the data and use them in the market planning process. Part of the analysis is weighing the importance of the trends so that the organisation can keep the trends in proper perspective (Jogaratnam & Law, 2006).

One of the most important tasks, especially in a small business such as a restaurant, is to assign responsibilities for the collection of data. However, in larger facilities, this is typically part of the job of a marketing manager while some organisations like the Georgia World Congress Centre in Atlanta appoint market research managers to play that role. Bob Southwest, former manager of the Houston Country Club, urged his secretarial staff to scan magazines for new menu ideas. Bar managers can look for lounge promotions. Dining room managers can study serving and promotional ideas. The staff then feeds ideas to the manager.

Using Information about the Marketing Environment

It is not sufficient to simply collect data about the environment. The information must be reliable, timely, and be used in decision-making. William S. Watson, Senior Vice President of Best Western Worldwide Marketing, offered some advice on this subject:

> As marketers, we are willing to make some intuitive leaps because of the creative aspects of our characters. Nevertheless, we need enough information to make reasonable decisions, and enough good data so that we can let our judgment move beyond the obvious, traditional interpretations we have learned as professionals. Researchers must put less emphasis on data and more on the interpretation of those data. They must work toward turning data into useful information. Collecting darn for its own value is like collecting stamps. It is a nice hobby but it does not deliver the mail.

The Business Unit Strategic Planning Process

Having examined the strategic planning tasks of company management, we can now examine the strategic planning tasks facing business unit managers. The business unit strategic planning process consists of eight steps. We shall examine these steps.

Business Mission

Each business unit needs to define its specific mission within the broader company mission. Thus, a strategic business unit (SBU) must define its various scopes more specifically, that is, its products, services, and applications, competence, market segments, vertical positioning, and geographical coverage. It must also define its specific goals and policies as a separate business (Dev et al., 2010).

SWOT Analysis

The overall evaluation of a company's strengths, weaknesses, opportunities, and threats is termed SWOT analysis. Strengths are the resources and capabilities within the organisation such as financial resources, competent staff, service culture, and technological resources which serve as a basis for continued success. Weaknesses however prevent an organisation from performing at an optimum level. These include depreciating equipment, obsolete equipment, poor service quality, and low efficiency. The internal analysis looks at the strengths and weaknesses of the company whiles the external analysis looks at opportunities and threats created by the environment. These comprise of opportunities and threats which are external factors that either offer opportunities for improvement and growth or inhibit progress. In strategic planning, the company matches its capabilities with the opportunities created by the environment and takes action to minimise environmental threats (Runtunuwu, 2020).

External Environmental Analysis (Opportunity and Threat Analysis)

The business manager now knows the parts of the environment to monitor if the business is to achieve its goals. In general, a business unit has to monitor key macroenvironmental forces (demographic-economic, technological, political-legal, and social-cultural) and significant microenvironmental forces (customers, competitors, distribution channels, suppliers) which will affect its ability to earn profits in this marketplace. The business unit should set up a marketing intelligence system to track trends and important developments. For each trend or development, management needs to identify the implied opportunities and threats.

Opportunities

A major purpose of environmental scanning is to discover new opportunities. We define a marketing opportunity as an area of need in which a company can perform profitably. Opportunities can be listed and classified according to their attractiveness and success probability. A company's success probability does not only depend on its business strengths matching the key success requirements for operating in the target market but also its ability to exceed the strengths of its competitors. The best-performing company will be the one that can generate the greatest customer value and sustain it over time.

Threats

Some developments in the external environment represent marketing threats. We define an environmental threat as a problem resulting from adverse trends or events that, in the absence of defensive marketing strategies, would result in a decline in sales or profits.

Threats should be classified according to their seriousness and probability of occurrence.

By assembling a picture of the major threats and opportunities facing a specific business unit, it is possible to characterise its overall attractiveness. Four outcomes are possible. An ideal business is high in major opportunities and low in major threats. A speculative business is high in both major opportunities and threats. A major business is low in opportunities and threats. Finally, a troubled business is low in opportunities and high in threats.

Internal Environmental Analysis (Strengths—Weaknesses Analysis)

It is one thing to discern attractive opportunities in the environment; it is another to have the necessary competencies to capitalise on those opportunities. Each business needs to evaluate its strengths and weaknesses periodically. Management or an outside consultant reviews the business' marketing, financial, manufacturing, and organisational competencies, each factor is rated as to whether it is a major strength, minor strength, neutral factor, minor weakness, or major weakness. A company with strong marketing capability would show up with the ten marketing factors all rated as major strengths.

In examining its pattern of strengths and weaknesses, clearly, the business does not have to correct all its weaknesses or gloat about all its strengths. The big question is whether the business should limit itself to those opportunities for which it now possesses the required strengths or should consider better opportunities.

Many hospitality experts believe that to compete effectively, companies such as hotels, resorts, and cruise lines will need seamless connectivity to computer reservation systems (CRS) and global distribution systems (GDS). If a hotel company wishes to increase its international business and its reservations through travel agents, the existence or development of such a system would surely be viewed as a strength (Azimovna et al., 2022).

Goal Formulation

After the business unit has defined its mission and examined its Strengths/Weaknesses/Opportunities/Threats (called SWOT analysis), it can proceed to develop specific objectives and goals for the planning period. This stage is called goal formulation.

Very few businesses pursue only one objective. Most business units pursue a combination of objectives, including profitability, sales growth, market-share improvement, risk containment, innovativeness, and reputation.

The business unit should strive to arrange its objectives hierarchically, from the most important to the least important. Where possible, objectives should be stated quantitatively. Objectives should be Specific, Measurable, Achievable, Realistic, and Time-bound (SMART). For instance, the objective "increase the return on investment (ROI)" is not as satisfactory as "increase ROI to 15 percent" or, even better, "increase ROI to 15 percent within two years." Managers use the term goals to describe objectives that are specific with respect to magnitude and time. Turning objectives into measurable goals facilitates management planning, implementation, and control. A business should set realistic goals. The levels should arise from an analysis of the business unit's opportunities and strengths, not from wishful thinking.

"The process involves determining the specific objectives of business or project speculation and identifying internal and external factors that support and which are able to help in achieving that goal" (Ritonga et al., 2018).

Finally, the company's objectives need to be consistent. Objectives are sometimes in a trade-off relationship. Here are some important trade-offs:

1. High-profit margins versus high market share.
2. Deep penetration of existing markets versus developing new markets.
3. Profit goals versus non-profit goals.
4. High growth versus low risk.

The hotel industry is faced with unique challenges concerning goal formulation and performance measurement, especially in situations where there are management agreements between hotel owners and hotel operating companies. Most industries, such as manufacturing, construction, or retailing, hire their management staff rather than contracting an independent operations management company.

Strategy Formulation

Goals indicate what a business unit wants to achieve while strategy indicates how to achieve it. In other words, goal is the destination while strategy is the journey. Every business must tailor a strategy for achieving its goals. Although we can list many types of strategies, Michael Porter has condensed them into three generic types, namely cost leadership, differentiation, and focus strategies that provide a good starting point for strategic thinking.

Overall Cost Leadership: Here the business works hard to achieve the lowest costs. This is usually achieved by offering a standardised product or service which is mass-produced so that the firm could achieve efficiency in production and reduction in costs of inputs through economies of scale. EasyJet, Europe's leading short-haul airline is famous for pursuing this strategy. The problem with this strategy is that other firms will usually emerge with still lower costs. The real key is for the firm to achieve the lowest costs among those competitors operating in the same industry.

Differentiation: Here the business concentrates on achieving superior performance in an important customer benefit area valued by a large part of the market. In reality, the relative importance of customer benefit areas tends to shift. As demographics and characteristics of market populations change, younger, active hotel guests may value the existence of a swimming pool, sauna, or gym, whereas older guests have been shown to place a high value on reliability and assurance of consistent hotel service. Intercontinental Hotels and Resorts, employs a differentiation strategy by offering exquisite facilities and superior service which is often in consonance with the unique local cultures of the location of the facilities. One major challenge with this strategy is that over time the differentiation becomes a norm as it is emulated by competitors and consumers no longer perceive it as a differentiation. For instance, Wi-Fi was initially deployed in hotels as a differentiation strategy but is now considered a necessity.

Focus: Here the business focuses on one or more narrow market segments rather than going after a large market. The firm gets to know the needs of these segments and pursues either cost leadership or a form of differentiation within the target segments. Some hospitality firms like Som Hotels in Spain are now offering women-only hotels. These hotels often attract businesswomen who travel alone and provide the requisite facilities and amenities that appeal to women.

According to Porter, those firms pursuing the same strategy directed to the same market or market segment constitute a strategic group. Porter suggests that firms that do not pursue a clear strategy become "stuck-in-the-middle." Where Stuck-in-the-middle is when organisations pursue all three strategies and are unable to achieve the full advantages of one of the strategies.

Programme Formulation

Once the business unit has developed its principal strategies, it must work out supporting programmes. Thus, if an upscale hotel has decided to attain service leadership, it must run recruitment programmes to attract the right employees, conduct training programmes, develop leading-edge products and amenities, motivate employees, and develop and communicate its service leadership, and so on (Ritonga et al., 2018).

Implementation

Even a clear strategy and well-thought-out supporting programmes may not be enough. The firm may fail at implementation. Employees in a company share a common way of behaving and thinking. They must understand and believe in the company's strategy. The company must communicate its strategy to the employees and make them understand their part in carrying it out. To implement a strategy, the firm must have the required resources, including employees with the necessary skills to carry out that strategy.

Feedback and Control

A firm needs to track results and monitor new developments in the environment as it implements its strategy. Some environments are fairly stable from year to year. Others evolve slowly in a fairly predictable way. Still, other environments change rapidly. The company can count on one thing: the environment will change. When that happens, the company will need to review and revise its implementation, programmes, strategies, or even objectives. Peter Ducker pointed out that it is more important to do the right thing (being effective) than to do things right (being efficient). Excellent companies excel at both.

Once an organisation starts losing its market position through failure to respond to a changing environment, it becomes increasingly harder to retrieve leadership. Organisations, especially large ones, have much inertia. Yet organisations can be changed through leadership, hopefully in advance of a crisis but certainly in the middle of one. The key to organisational health is the organisation's willingness to examine the changing environment and to adopt appropriate goals and behaviours. Adaptable organisations monitor the environment continuously and attempt flexible strategic planning to maintain a fit with the evoking environment (Kim et al., 2018).

Summary

In this chapter, we discussed the marketing environment within which businesses and for that matter, tourism and hospitality businesses fall. We also went further to examine how managers can use information about the marketing environment to leverage business proceeds. We then discussed how to design business portfolios and growth strategies. The steps involved in the business strategy planning process were also explained.

Activity

1. Define the marketing environment of a business in the hospitality and tourism industry.
2. Outline how elements in the marketing environment affect tourism and hospitality businesses in Africa.
3. Briefly explain the microevironmental factors and how they affect a company's ability to create value and satisfy its customers.
4. Identify the major macroevironmental factors and how they can affect companies in the hospitality and tourism industry.
5. Assuming you are an executive in the hospitality and tourism industry, how will you respond to the threat of macroevironmental factors?

References

Azimovna, M. S., Shokhrukhovich, U. F., & Sodirovich, U. B. (2022). Analysis of the market of tourist products of the Samarkand region. *barqarorlik va yetakchi tadqiqotlar onlayn ilmiy jurnali*, *2*(4), 422–427.

Banna, H., Mia, M. A., & Rana, M. S. (2021). The role of banking and financial institutions in the tourism and hospitality industry in Bangladesh. In *Tourism in Bangladesh: Investment and development perspectives* (pp. 111–123).

Dev, C. S., Buschman, J. D., & Bowen, J. T. (2010). Hospitality marketing: A retrospective analysis (1960–2010) and predictions (2010–2020). *Cornell Hospitality Quarterly*, *51*(4), 459–469.

Hassanien, A., Dale, C., Clarke, A., & Herriott, M. W. (2010). *Hospitality business development*. Routledge.

Jogaratnam, G., & Law, R. (2006). Environmental scanning and information source utilization: Exploring the behavior of Hong Kong hotel and tourism executives. *Journal of Hospitality & Tourism Research*, *30*(2), 170–190.

Kim, M., Lee, S. K., & Roehl, W. S. (2018). Competitive price interactions and strategic responses in the lodging market. *Tourism Management*, *68*, 210–219.

Kracht, J., & Wang, Y. (2010). Examining the tourism distribution channel: Evolution and transformation. *International Journal of Contemporary Hospitality Management*. *22*(5), 736–757.

Ritonga, H. M., Setiawan, N., El Fikri, M., Pramono, C., Ritonga, M., Hakim, T., & Nasution, M. D. T. P. (2018). Rural tourism marketing strategy and swot analysis: A case study of Bandar PasirMandoge sub-district in North Sumatera. *International Journal of Civil Engineering and Technology*, *9*(9), 1617–1631.

Runtunuwu, P. C. H. (2020). Tourism sector development strategy in North Maluku: A case study of Tidore Islands. *International Journal of Tourism and Hospitality in Asia Pasific (IJTHAP)*, *3*(3), 1–8.

Sijabat, R. (2019). Macro-environment analysis of the tourism industry in Indonesia: Findings from the PEST analysis. *Innovative Issues and Approaches in Social Sciences*, *12*(3), 101.

Chapter 7

Marketing Information Systems and Marketing Research

Chapter Outline

- Introduction
- The Marketing Information System
- Assessing Information Needs
- Developing Information
- Internal Records
- Marketing Intelligence
- Internal Sources of
- Marketing Intelligence
- Commercial Sources of Marketing Information
- Marketing Research
- Marketing Research Process
- Defining the Problem and Research Objectives
- Developing the Research Plan
- Planning Primary Data Collection
- Implementing the Research Plan
- Interpreting and Reporting the Findings

DOI: 10.4324/9781032688497-7

Objectives

At the end of this chapter, you should be able to:

- Explain the concept of marketing information systems.
- Identify the different kinds of information a hospitality or tourism company might use.
- Identify sources of information.
- Outline the marketing research process, including defining the problem and research objectives, developing the research plan, implementing the research plan, and interpreting and reporting the findings.

Introduction

In this chapter, we examine marketing information systems (MIS) and marketing research as relevant components of hospitality and tourism marketing. In carrying out marketing analysis, planning, implementation, and control, marketing managers need information at almost every turn. They need information about customers, competitors, suppliers, and other forces in the marketplace. One marketing executive puts it this way: "To manage a business well is to manage its future, and to manage the future is to manage information."

During the past century, most hotels and restaurants were independently owned or they were part of a small regional chain. Managers obtained information by being around people, observing them, and asking questions. During this century, many factors have increased the need for more and better information. As companies become bigger and expand into international markets, they need information on those markets. As companies become more selective, they need better information about how buyers respond to different products and appeals. As companies use more complex marketing approaches and face intensified competition, they need information on the effectiveness of their marketing tools. Finally, in today's rapidly changing business environments, managers need up-to-date information to make timely decisions.

"Technology revolutionises the tourism industry and determines the strategy and competitiveness of tourism organisations and destinations" (Buhalis, 2020).

The supply of information has also increased greatly. Using improved computer systems and other technologies, companies can provide more information. In fact, today's managers sometimes receive too much information. For example, one study found that with many companies offering data and information available through supermarket scanners, a packaged-goods brand manager is bombarded with one million to one billion new numbers each week. As Naisbitt (1994) points out, "Running of our information is not a problem, but drowning in it is." Yet managers frequently complain that they lack enough information on the right kind and accumulate too much of the wrong kind. They also complain that marketing information is so widely spread throughout the company that it takes great effort to locate even simple facts. Subordinates may withhold information they believe will reflect badly on their performance. Important information often arrives too late to be useful or is not accurate. Marketing managers need precise and timely information. Many companies are now studying their managers' information needs and designing systems to meet those needs (Law et al., 2020).

The Marketing Information System

A marketing information system (MIS) comprises personnel, tools, and processes to collect, classify, analyse, and deliver accurate information to marketing decision-makers promptly. The MIS begins and ends with marketing managers, but all managers throughout the organisation should be involved in the MIS. First, it interacts with managers to assess their information needs. Next, it develops the needed information from internal company records, marketing intelligence activities, and the marketing research process. Information analysts process information to make it more useful. Finally, the MIS distributes information to managers in the right form and at the right time to help in marketing planning, implementation, and control. We now take a closer look at the functions of a company's marketing information systems.

Assessing Information Needs

A good MIS balances information that managers would like to have against that which they really need and is feasible to obtain. The company begins by interviewing managers to determine their information needs. For example, Mrs. Field's Cookies provides their managers with sales forecasts with updates each hour. When sales are falling behind, the computer suggests appropriate sales techniques for revamping sales.

Some managers will ask for whatever information they can get without thinking carefully about its cost or usefulness. Too much information can be as harmful as too little. Other busy managers may fail to ask for tailings they need to know, or managers may not ask for some types of information they should have. (For example, managers need to anticipate new products to be offered by their competitors. However, companies withhold information from their competitors. When KFC was developing its "Chicken Little" sandwich, only a few corporate managers knew of the project. KFC had developed ingredient specifications for preparing the sandwich, and its suppliers had to sign secrecy agreements. KFC did not want competitors to learn about the new product before its test marketing. Yet a competitor with a good MIS system might have picked up clues in advance about KFC's plans. For instance, they may have heard a bread supplier passing comment about KFC's orders for small hamburger-style buns. They may have heard an executive stating how KFC would be strengthening its lunch business. Even with secret agreements, news inadvertently leaks out. Managers who keep their ears to the ground can pick up information on competitive moves using legal and ethical sources of information such as speeches by company executives and trade publications.

Developing Information

Information needed by marketing managers can be obtained from internal company records, marketing intelligence, and marketing research. The information analysis system processes this information and presents it in a form that is useful to managers.

Internal Records

Most marketing managers use internal records and reports regularly for day-to-day planning, implementation, and control decisions. Internal records information consists of information gathered from sources within the company to evaluate marketing performance and detect marketing problems and opportunities. The company's accounting department develops financial statements and keeps detailed records of sales, costs, and cash flows. Daily reports from restaurants can include total sales, sales by a service person, and sales by menu items, guest counts, and average check. Hotel daily reports can include occupancy, number of guests, total revenue, average daily rate (ADR), revenue per available room (REVPAR), no-shows, and groups in the house.

Many companies use internal records to develop comprehensive internal databases, and computerised information obtained from varied data sources within the company. Marketing managers can readily access and work with information in the database to identify marketing opportunities and problems, plan programmes, and evaluate performance. Information derived from the database could also be analysed and employed for customer relationship management (CRM) including leading sales and marketing campaigns and increasing customer engagement.

Internal databases usually can be accessed more quickly and cheaply than other information sources, but there are certain challenges with their use. Because the data were collected for other purposes, it may not be complete or suitable for making marketing decisions. For example, information from the point of sales in a restaurant software system used for ordering from the kitchen and the development of guest checks can be used for server evaluation, product evaluation, and to collect information on individual customers. Because data ages quickly a major effort is required to keep the database current. In addition, large companies produce lots of information, and keeping track of all of such information is difficult. The database information must be well integrated and should be readily accessible through user-friendly interfaces so that managers can find them useful. This also enables managers to be effective in the operation of their organisations (Shkeer & Awang, 2019).

Guest History Information: The single most important element in any hospitality marketing information system is the existence of a process for capturing and using guest information. Guest information is vital to improving service, creating effective advertising and sales promotion programmes, developing new products, improving existing products, developing marketing and sales plans, and developing and using an effective revenue management programme. Unfortunately, many hospitality firms have only a vague idea of who their guests are.

Specific guest information needs may include name, marital status, address, emergency contact, length-of-stay, allergies, transactions history, guest actions, and preferences. At first appearance, this list undoubtedly seems overbearing and unduly inquisitive. The fact is that hospitality companies increasingly collect and use this type of information. Obviously, hospitality businesses including hotels, resorts, restaurants, and cruise lines must be very careful not to infringe on the privacy rights of guests or to disturb them. A significant amount of this information could be derived from internal records. This requires interfacing with other departments, such as rooms division, marketing, and accounting.

Automated Systems: The decrease in costs, inspite of the increasing capacity of automated guest history systems will compel hotels to create close relationships with their customers once again. Obviously, any hospitality company, such as a large cruise line, must utilise an automated system. A variety of systems are available and should be examined carefully and tested before purchasing. Remember that an automated guest information system is part of larger systems such as database marketing and yield/revenue management.

An automated guest information system can be of great benefit to the sales force. Salespeople can pull guest histories by a specific geographic area, such as a city. This information can greatly assist in a sales blitz by identifying frequent guests who can receive top priority in the blitz. The guest information system can also identify former frequent guests who are no longer patronising the hotel. Salespeople will want to call on these former clients to see if they can regain their business.

Disguised Shoppers: Hospitality companies often hire disguised or mystery shoppers to pose as customers and report back on their experiences. Industry professionals are recruited to evaluate the products and services offered by hospitality establishments. Mystery shoppers sample the actual guest experience and engage in normal guest–staff encounters and interactions. Through this, they are able to evaluate the quality of the services and facilities and report to management. The report serves as a basis for improving the overall performance of the business. The managers of Ruby's, a chain of restaurants based in California, use shoppers to alert managers and employees to pay attention to important areas of their operations. Employees realise that they may be "shopped."

A mystery shopper works best if there is a possibility for recognition and reward for good job performance. This is the concept of positive reinforcement. If employees feel that the only purpose of a disguised shopper programme is to report poor service and reprimand them, the programme will not fulfil its objective.

Company Records: One of the most misused sources of information is company records. Marketing managers should take advantage of the information that is currently being generated by various departments. Guest history and history on potential corporate clients are also vital information.

Point-of-Sale Information: For restaurants, the point-of-sale (POS) register undoubtedly offers opportunities to compile and distribute information through a computer, information that is currently entered into reports manually. A POS system automatically collects information about individual restaurant patrons where credit cards are used.

Some observers of the fast-food industry believe that future POS systems will use expert systems that employ robotics and computers using artificial intelligence. One possible scenario is the "computaburger." Data concerning customer preference, order size, and volume will be taken from a POS machine and provided to an expert system. The expert system will then predict and possibly even order a volume of hamburgers and the accompanying condiments for specific times each day. Due to the COVID-19 pandemic, some hospitality businesses had to deploy contactless technology including robotics. For instance, some Royal Caribbean Cruise ships have Robotic bars which sell robot made cocktails. People place their orders by scanning their IDs.

The casino industry has also displayed a high interest in POS systems and their increasing sophistication. Some slot machines are now capable of recording the number of plays and the win/loss record of frequent players who activate the machines through the use of a magnetic card. The player receives points based on the amount of play, and the casino is able to track the playing habits of players using slot club cards.

Systems are also in place in most casinos to track players who are brought to the casino by junket reps. Tracking these players is the responsibility of the pit boss in each gaming area, such as blackjack.

This information can be obtained from annual reports, financial analyses of public companies, articles on the company, and by talking to company employees. In addition to detailed information concerning prospects, sales force members are required to be regular readers of the business press, such as the *Wall Street Journal* and the *New York Times.*

Marketing Intelligence

Marketing intelligence includes everyday information about developments in the marketing environment that helps managers prepare and adjust marketing plans and short-run tactics. Marketing intelligence systems determine the needed intelligence and collect and deliver it in a useful format to marketing managers.

Dynamic, volatile, and time-sensitive industries, such as tourism, travel, and hospitality require agility and market intelligence to create value and achieve competitive advantage (Stylos et al., 2021).

Internal Sources of Marketing Intelligence

Marketing intelligence can be gathered by a company's executives, front-desk staff, service staff, purchasing agents, and sales force. Employees, unfortunately, are often too busy to pass on important information to management. The company must inform them about their role as intelligence gatherers and train them on how to spot and report new developments. Managers should debrief contact personnel regularly.

Commercial Sources of Marketing Information

Companies can also purchase information from external suppliers. One such source of information is a system called *Dialogue* which provides access to over 350 databases. At the click of a button a managers can retrieve information on new products and locations, industry trends and projections, press releases, and the detailed finances of public and private businesses. Today there are thousands of online databases of information services. For example, the Ad track online database tracks all the advertisements of a quarter-page or larger from 150 major consumer and business publications. Companies can use these data to assess advertising strategies and styles, shares of advertising space, media use, and ad budgets.

Associations sometimes collect data from their member companies, compile it, and make it available to members at a reasonable fee. Information of this nature should be used with care as they could be misleading because member companies frequently provide incorrect data or may refuse to contribute any statistics if they have a dominant market share.

Marketing Research

Marketing research is a process that identifies and defines marketing opportunities and problems, monitors, and evaluates marketing actions and performance, and communicates the findings and implications to management.

Marketing researchers engage in a wide variety of activities. Their ten most common activities are the measurement of market potentials, market share analysis, and the determination of market characteristics, sales analysis, and studies of business trends, short-range forecasting, competitive product studies, long-range forecasting, marketing information systems studies, and testing of existing products.

A company can conduct marketing research by employing its own researchers or hiring external researchers. Most large companies—in fact, more than 73 percent—have marketing research departments that conduct their marketing research. But even companies with research departments also hire outside firms to do fieldworks, data analysis, and other special tasks.

Marketing Research Process

The marketing research process consists of four major steps namely defining the problem and research objectives, developing the research plan, implementing the research plan, and interpreting and reporting the findings (Wilson et al., 2010).

Defining the Problem and Research Objectives

Managers must work closely with marketing researchers to define the problem and the research objectives. The manager best understands the problem or decision for which information is needed, whiles the researcher best understands the research processes and methods to use to obtain information.

Managers must know enough about marketing research to interpret the findings. If they know little about marketing research, they may accept the wrong information, draw the wrong conclusions, or request much more information than they need. Marketing researchers can help the manager define the problem, interpret the results, and use the findings correctly.

In one case a restaurant manager hired a researcher to determine the level of consumer awareness of the restaurant. The manager felt that lack of awareness explained low patronage. The researcher found, to the contrary, that many people were aware of the restaurant but thought of it as a special occasion rather than an everyday restaurant. The manager had misdefined the problem and the research objective.

Assuming that the problem is well defined, the manager and researcher must set research objectives. A marketing research project can have one of the three types of objectives: exploratory research, to gather preliminary information that will help define the problem and suggest hypotheses; descriptive research, to describe the size, nature, and composition of the market; and causal research, to test hypotheses about cause-and-effect relationships. Managers often start with exploratory research and later follow with descriptive and/or causal research.

Developing the Research Plan

The second step in marketing research calls for determining the needed information and developing a data collection plan.

Determining Specific Information Needs: Research objectives must be translated into specific information needs. When Marriott decided to research a new, lower-priced hotel system, it had two goals: to pull travellers away from competitors and to minimise cannibalisation of its existing hotels. This research might seek to answer the following specific research questions:

What features should the hotel offer?
How should the new hotels be priced?
Where should the hotels be located? Can they safely be located near existing Marriott hotels without incurring cannibalisation?
What are the probable sales and profits?

Gathering Secondary Information: To meet a manager's information needs, researchers can gather secondary data, primary data, or both. Secondary data consist of information already in existence somewhere, having been collected for another purpose. Primary data consist of information collected for the specific purpose at hand.

Researchers usually start by gathering secondary data. Secondary data are usually obtained more quickly and at a lower cost than primary data.

Planning Primary Data Collection

The major advantage of survey research is its flexibility. It can be used to obtain different kinds of information in many different marketing situations.

Depending on the survey design, it may also provide information more quickly and at a lower cost than can be obtained through observational or experimental research.

Survey research also has some limitations. Sometimes people are unable to answer survey questions because they cannot remember or never thought about what they do and why. Or they may be reluctant to answer questions asked by unknown interviewers about things that they consider private. Busy people may not want to take the time. Respondents may answer survey questions even when they do not know the answer to appear smart or well-informed. Or they may try to help the interviewer by giving pleasing answers. Careful survey design can help minimise these problems.

Although observation is best suited for exploratory research and surveys for descriptive research, experimental research is best suited for gathering data on cause–effect scenarios. Experiments involve selecting matched groups of subjects, giving them different treatments, controlling unrelated factors, and checking for differences in group responses.

Researchers at Arby's might use experiments before adding a new sandwich to the menu to answer the following research questions:

By how much will the new sandwich increase Arby's sales?
How will the new sandwich affect the sales of other menu items?
Which advertising approach would have the greatest effect on sales of the sandwich?
How would different prices affect the sales of the product?
Should the new item be targeted towards adults, children, or both?

Implementing the Research Plan

The researcher puts the marketing research plan into action by collecting, processing, and analysing the data. Data collection can be done by the company's marketing research staff, which affords the company greater control of the collection process and data quality. It could also be undertaken by outside firms. Outside firms that specialise in data collection can often do the job more quickly at a lower cost.

The data-collection phase of the marketing research process is generally the most expensive and the most subject to error. The researcher should watch the fieldwork closely to ensure that the plan is implemented correctly and to guard against problems with contacting respondents who refuse to cooperate or who give biased or dishonest answers, and interviewers who make mistakes or take shortcuts.

The collected data must be processed and analysed to pull out important information and findings. Data from questionnaires are checked for accuracy and completeness and coded for computer analysis. The researcher applies standard computer programmes to prepare tabulations of results and to compute averages and other measures for the major variables.

Interpreting and Reporting the Findings

At the final stage, the researcher must now interpret the findings, draw conclusions, and report the findings to management. The researcher should avoid overwhelming managers with numbers, complex statistical techniques, and focus. Instead, management desires major findings that will be useful in decision-making.

Interpretation should not be left entirely to the researcher. Findings can be interpreted in different ways, and discussions between researchers and managers will help point to the best interpretations. The manager should also confirm that the research project was executed properly. After reviewing the findings, the manager may raise additional questions that can be answered with research data. Researchers should make the data available to marketing managers so that they can perform further analyses and test relationships on their own.

Interpretation is an important phase of the marketing process. The best research is meaningless if a manager blindly accepts wrong interpretations. Similarly, managers may have biased interpretations. They sometimes accept research results that show what they expected and reject those that did not provide expected or hoped-for answers. Thus, managers and researchers must work closely together when interpreting research results. Both share responsibility for the research process and resulting decisions.

Managers need to remember that research is a process, and that the researcher must proceed through all steps of the process (Esfehani & Walters, 2018).

Summary

In this chapter, we studied the concept of marketing information systems where we identified the different kinds of information the company might use. We also outlined the marketing research process, including defining the problem and research objectives, developing the research plan, implementing the research plan, and interpreting and reporting the findings.

Activity

1. Drawing on relevant hospitality and tourism examples, discuss the marketing research process.
2. What is marketing information systems and how can managers in the hospitality industry use it to create value for customers and increase profits?
3. Explain marketing intelligence and identify the different sources of it in a hospitality or tourism entity.

References

Buhalis, D. (2020). Technology in tourism-from information communication technologies to eTourism and smart tourism towards ambient intelligence tourism: A perspective article. *Tourism Review, 75*(1), 267–272.

Esfehani, M. H., & Walters, T. (2018). Lost in translation? Cross-language thematic analysis in tourism and hospitality research. *International Journal of Contemporary Hospitality Management, 30*(11), 3158–3174.

Law, R., Leung, D., & Chan, I. C. C. (2020). Progression and development of information and communication technology research in hospitality and tourism: A state-of-the-art review. *International Journal of Contemporary Hospitality Management, 32*(2), 511–534.

Naisbitt, J. (1994). Global paradox: The bigger the world economy, the more powerful its smallest players. *Journal of Leisure Research, 26*(4), 406.

Shkeer, A. S., & Awang, Z. (2019). The impact of marketing information system components on organizational decision making: A case of Jordanian five star hotels. *International Review of Management and Marketing, 9*(6), 197.

Stylos, N., Zwiegelaar, J., & Buhalis, D. (2021). Big data empowered agility for dynamic, volatile, and time-sensitive service industries: The case of tourism sector. *International Journal of Contemporary Hospitality Management, 33*(3),1015–1036.

Wilson, A., Johns, R., Miller, K. W., & Pentecost, R. (2010). Marketing research: An integrated approach: Solving business problems: The value of marketing research. Pearson.

Chapter 8

Consumer Behaviour

Chapter Outline

- Introduction
- Personality and Self-Concept
- Psychological Factors
- Motivation
- Perception
- Learning
- Beliefs and Attitudes
- The Buyer Decision Process
- Need Recognition
- Information Search
- Evaluation of Alternatives
- Purchase Decision
- Post-purchase Behaviour

Objectives

After reading this chapter, you should be able to:

- Outline the major characteristics affecting consumer behaviour.
- List some of the specific cultural, social, personal, and psychological factors that influence consumer behaviour.
- Explain the factors influencing consumer buying behaviour.
- Explain the buyer decision process.

DOI: 10.4324/9781032688497-8

Introduction

Marketers must exercise care in analysing consumer behaviour. Consumers often turn down what appears to be a winning offer. As soon as managers believe that they understand their customers, buyer decisions are made that appear to be irrational. But what looks like irrational behaviour to a manager is completely rational to the consumer. Buying behaviour is never simple. It is affected by diverse factors, yet understanding it is an essential task of the marketing management. This section, takes a look at consumer markets and consumer buying behaviour.

Personality and Self-Concept

Personality influences one's buying behaviour. By personality, we mean an individual's distinguishing psychological characteristics that lead to relatively consistent and enduring responses to the environment. Personality can be useful in analysing consumer behaviour regarding some product or brand choices. For example, a beer company may discover that heavy beer drinkers tend to rank high in sociability and aggressiveness. This information can be used to establish a brand image for the beer and to suggest the type of people to show in an advertisement. Kim et al. (2020) believes that personality can affect consumers' choice of all products and services including tourist attractions.

Many marketers use a concept related to personality: a person's self-concept (also called self-image). Each of us has a complex mental self-picture, and our behaviour tends to be consistent with that self-image. A person who perceives himself as outgoing and active is unlikely to purchase a cruise vacation if his perception of a cruise is one of elderly persons lying on lounge chairs. Such a person is more likely to select a scuba-diving or skiing vacation. The cruise line industry has been quite successful in changing its "geriatric" image and now attracts outgoing and active consumers.

The role of self-concept obviously has a strong bearing on the selection of recreational pursuits, including golf, sailing, dirt bike riding, fishing, and hunting. Anyone who enjoys boating will attest to the fact that there is a difference between boaters who use sails and those who use engines. Yachters/sailboaters refer to those who use engines as "stink potters." Stink potters think of the sailing crowd as stuffy, pretentious, and generally not much fun.

Psychological Factors

A person's buying choices are also influenced by four major psychological factors namely motivation, perception, learning, and beliefs and attitudes.

Motivation

People tend to have many needs at any given time. Some are biological, arising from hunger, thirst, and discomfort. Others are psychological, arising from states of tension, such as the need for recognition, esteem, or belonging. Most of these needs are not strong enough to motivate a person to act at a given point in time. A need becomes a motive when it is aroused to a sufficient level of intensity. Creating a tense state causes the person to act to release the tension. Psychologists have developed theories of human motivation. Two of the most popular theories, Abraham Maslow's Hiearchy of Needs, and Herzberg's Two Factor Theory, have quite different meanings for consumer analysis and marketing.

Maslow's Theory of Motivation: Abraham Maslow sought to explain why people are driven by particular needs at particular times. Why does one person spend much time and energy on personal safety and another on gaining the esteem of others? Maslow answers that human needs are arranged in a hierarchy, from the most pressing to the least pressing. Maslow's hierarchy of needs, in order of importance, are physiological needs, safety needs, social needs, esteem needs, and self-actualisation needs. A person tries to satisfy the most important need first. When that important need is satisfied, it will stop being a motivator, and the person will then try to satisfy the next most important need.

Herzberg's Theory: Frederick Herzberg developed a two-factor theory that distinguishes dissatisfiers (factors that cause dissatisfaction) and satisfiers (factors that cause satisfaction). According to Herzberg, the absence of dissatisfiers is not enough; satisfiers must be actively present to motivate a purchase. For example, a computer that does not come with a warranty would be a dissatisfier. Yet the presence of a product warranty would not act as a satisfier or motivator of a purchase, because it is not a source of intrinsic satisfaction with the computer. Ease of use would be a satisfier.

Herzberg's theory has two implications. First, hospitality service providers should do their best to avoid dissatisfiers (for example, dirty surroundings or a poor service quality). Although these things will not sell a product, they might easily "unsell" it. Second, the manufacturer should identify the major satisfiers or motivators of purchase in the market and then supply them. These satisfiers will make a major difference as to which brand the customer buys.

Perception

How a motivated person acts is influenced by his or her perception of the situation. In the same situation, two people with the same motivation may act quite differently based on how they perceive conditions. One person may perceive the waiters at T.G.I. Friday's as casual and unsophisticated, while another person may view them as spontaneous with cheerful personalities. Friday's is targeting those in the second group.

Why do people have different perceptions of the same situation? All of us experience a stimulus through the flow of information through our five senses: sight, hearing, smell, touch, and taste. However, each of us receives, organises, and interprets this sensory information differently. Perception is the process by which an individual selects, organises, and interprets information to create a meaningful picture of the world.

The key word in the definition of perception is "individual." One person might perceive a fast-talking salesperson as aggressive and insincere but another person might perceive the same person as intelligent and helpful. People can emerge with different perceptions of the same object because of three perceptual processes: selective attention, selective distortion, and selective retention.

Selective Attention: People are exposed to a tremendous number of daily stimuli: the average person may be exposed to over 1,500 advertisements a day because a person cannot possibly attend to all of these, most stimuli will be screened out—a process called selective attention. Selective attention means that marketers have to work hard to attract consumers' attention. The real challenge is to explain which stimuli people will notice. Here are Mime's findings:

People are more likely to notice stimuli that relate to a current need. A person who is motivated to buy pizza will notice pizza ads; they will probably not notice Cheeseburger ads. People are more likely to notice stimuli they anticipate. You are more likely to notice the chicken burgers in a fast-food restaurant than pizza because you do not expect the restaurant to offer pizzas.

People are more likely to notice stimuli whose deviations are larger than the normal size of the stimuli. You are more likely to notice an advert offering $10 off the list price of a burger than one offering $2 off.

Selective Distortion: Stimuli do not always come across in the way the senders intended. Selective distortion is the tendency for people to distort information and interpret it in a way that will fit their previously held notions. Unfortunately, there is not much that marketers can do about selective distortion.

People with selective retention tend to forget much of what that they learn but will retain information that supports their attitudes and beliefs. Because of selective retention, we are likely to remember good points mentioned about competing products. Selective retention explains why marketers use drama and repetition in sending messages to their target market.

Learning

People learn through their actions. Learning describes changes in an individual's behaviour arising from experience. Most human behaviour is learned. Learning theorists say that learning occurs through the interplay of drivers, stimuli, cues, responses, and reinforcement.

When consumers experience a product, they learn about it. Members of the venue committee for a convention often sample the services of competing hotels. They eat in the restaurants, experience the friendliness and professionalism of the staff, and assess the hotel's facilities. Based on what they have learned, a hotel is selected to host the convention. During the convention, they experience the hotel once again. Based on their experience and that of the attendees, they will either be satisfied or dissatisfied.

Hotels should help guests to learn about the quality of their facilities and services by providing the necessary information through diverse platforms. Luxury hotels organise tours for first-time guests and tour operators and inform them of the services they offer. They equally keep repeat guests updated on the hotel's services.

Beliefs and Attitudes

Through acting and learning, people acquire beliefs and attitudes, which, in turn, influence their buying behaviour. A belief is a subjective opinion that a person holds about something. A customer may believe that Adam's Mark Hotels have the best facilities and most professional staff among hotels in that category. These beliefs may be based on real knowledge, opinion, or faith. They may or may not carry an emotional charge.

Marketers are interested in the beliefs that people have about specific products and services because they can build their brands around such beliefs. Beliefs reinforce product and brand images. Also, people act on their beliefs. Therefore, if such beliefs are unfounded, they deter purchases and marketers should launch a campaign to change them.

Unfounded consumer beliefs can severely affect the revenue and even the survival of hospitality and travel companies. These beliefs might include the following:

- A particular hamburger chain served ground kangaroo meal.
- A particular hotel served as Mafia headquarters.
- A particular airline has a poor maintenance culture.
- A particular country has unhealthy food-handling standards.

People have attitudes towards almost everything: religion, politics, clothes, music, and food. An attitude describes a person's relatively consistent evaluations, feelings, and tendencies towards an object or an idea. People's attitude will determine whether they like or dislike things and whether they will patronise it or not. For example, many people who have developed the attitude that eating healthy food is important, perceive chicken as a healthier alternative to beef and pork. As a result, the per capita consumption of chicken has increased in recent years, leading the American Beef Council and National Pork Producers Council to try to change consumer attitudes that beef and pork are unhealthy. The National Pork Producers Council promotes pork as "the other white meat," trying to associate pork with chicken. Companies can benefit by researching attitudes towards their products. Understanding attitudes and beliefs are the first steps towards changing or reinforcing them.

Attitudes are very difficult to change. A person's attitudes fit into a pattern, and changing one attitude may require making many difficult adjustments. It is easier for a company to create products that are compatible with existing attitudes than to change the attitudes towards their products. There are exceptions, of course, where the high cost of trying to change attitudes may pay off. There is a saying among restaurateurs that a restaurant is only as good as the last meal served. Attitudes explain in part why this is true. A customer who has paid repeat visits to a restaurant and during one of such visits receives a bad meal may begin to believe that it is impossible to count on having a good meal at that restaurant. The customer's attitude towards the restaurant begins to change. If this customer again receives a bad meal, negative attitudes may be permanently fixed and this will prevent a future return to the restaurant. Serving a poor meal to first-time customers can be disastrous because the restaurant will not have another chance to change that negative first impression the customer forms about the restaurant. Customers develop an immediate negative attitude that prevents them from returning.

Attitudes that people develop as children often influence their purchases as adults. Children may retain negative attitudes towards certain vegetables, people, and possibly places. Chances are equally good that they may retain positive images about certain brands like McDonald's and DisneyWorld due to a positive childhood experience with those brands. Hospitality and travel companies are particularly subject to lifelong consumer attitudes that result from positive or negative childhood experiences. Harsh words from the manager of a miniature golf course or air sickness on a commercial flight in which the flight attendant showed little sympathy, are negative attitude-building experiences.

Once negative attitudes are developed, they are hard to change. New restaurant owners often want quick cash flow and sometimes start without excellent quality. A new restaurateur complained that customers are fickle. When his restaurant first opened, there were lines of people waiting to be served. A few months later, he had plenty of empty seats every night. Obviously, he had not satisfied his first guests. Even though he may have subsequently corrected his early mistakes, his original customers had been disappointed, were not returning, and probably were reporting their negative experiences to their friends and relatives.

We can now appreciate the many individual characteristics and forces influencing consumer behaviour. Consumer choice is the result of a complex interplay of cultural, social, personal, and psychological factors. Many of these cannot be influenced by the marketer; however, they help the marketer to better understand customers' reactions and behaviours.

The Buyer Decision Process

Let us now look at how consumers make their buying decisions. The decision process consists of five stages, namely need-recognition, information search, evaluation of alternatives, purchase decision, and post-purchase behaviour. This model emphasises that the buying process starts long before the actual purchase and continues long after the actual purchase. It encourages the marketer to focus on the entire buying process rather than just the purchase decision. The model shows that consumers pass through five stages with every purchase they make. However, with some more routine purchases, consumers skip or reverse some of these stages. A customer in a bar purchasing a glass of beer may go right to the purchase decision, skipping information search and evaluation. This is referred to as an automatic response loop. The dream of every marketer is to have customers develop an automatic response to purchase their products. However, this does not typically happen.

Need Recognition

The buying process commences when the buyer recognises a problem or need. The buyer senses a difference between his or her actual state and the desired state. The need can be triggered by internal stimuli. From previous experience, the person has learned how to cope with this need and is motivated towards objects that he or she knows will satisfy it.

Needs can also be triggered by external stimuli. For instance, Rosemary passes a restaurant, and the aroma of freshly baked bread stimulates her hunger; she has lunch with a friend who just came back from Bali and raves about her trip, or she watches a television commercial on a Hyatt resort. All these stimuli can lead her to recognise a problem or need such as a need to escape by taking a vacation.

At this stage, marketers must determine the factors and situations that trigger consumer need or problem recognition. They should research consumers to find out what kinds of needs or problems led them to purchase a particular item, what brought about these needs, and how they led the consumers to choose this particular product.

Rosemary might have mentioned that she passed a card shop and noticed birthday cards, which reminded her that her boyfriend's birthday was approaching. She knew he liked German food, so she decided to take him to a German restaurant on his birthday.

By gathering such information, marketers can identify stimuli that most often trigger interest in their products and develop marketing programmes that involve these stimuli. Marketers can also show how their products are a solution to a problem. For example, T.G.I. Fridays advertised its gift certificates as a solution to Christmas shopping. Friday's food and atmosphere attract a broad range of people; the gift certificates are easy to buy, customers can avoid the hustle and bustle of going through crowded shopping centres; and they can be bought in denominations that fit planned expenditures. Friday's promoted gift certificates as a solution to common problems experienced before Christmas.

When choosing a hotel, business travellers want a hotel that provides them with the resources they need to work efficiently. They want a hotel with competent staff on duty, free and fast WiFi more than they want personalised services and fancy surroundings. Unfortunately, some hotels seem to confuse product opulence with providing features that will be beneficial to the business traveller. They provide lobbies and restaurants but give business travellers rooms that are not equipped as an office away from home, thereby failing to meet the needs of this important market.

The Travel Industry Association of America found that families with both heads of household employees were finding it difficult to find a week when everyone was free. As a result, this segment needed three- and four-day getaways that could be booked at the last minute, because 42 percent of this group makes plans within two weeks of the actual vacation. These examples show that businesses must understand the needs of their customers and how these needs are translated into product features that will appeal to the market.

Information Search

Consumers who have been prompted by either an internal or external stimulus may or may not search for more information. If the consumer's drive is strong and a satisfying product is near at hand, the consumer is likely to buy it at that moment. If not, the consumer may simply store the need in his/her memory and search for relevant information.

The extent of the search by the consumer will depend on the strength of the drive, the amount of initial information, the ease of obtaining more information, the value placed on additional information, and the satisfaction one gets from searching.

Rosemary asked several of her friends if they knew of a good German restaurant. Then she scanned a city magazine restaurant listing. Finally, she searched online to see if she could find additional German restaurants. As a result of her search, Rosemary identified three German restaurants. She then tried to find friends and acquaintances who had been to one or more of the restaurants to get their impressions. She also looked in the Zagat Restaurant Guide for her city to see how the restaurants were rated.

The consumer can obtain information from several sources. These include:

- Personal sources: family, friends, neighbours, acquaintances.
- Commercial sources: advertising, salespeople, dealers, packaging, displays.
- Public sources: restaurant reviews, editorials in the travel section consumer-rating organisations, online reviews like TripAdvisor.

With hospitality and travel products, personal and public sources of information are more important than advertisements. This is because a customer does not know what they are going to receive until they have received it. A customer cannot try out an intangible product before they purchase it. For example, people may hear of a restaurant through advertising but ask their friends about the restaurant before they try it. Responses from personal sources have more impact than advertising because they are perceived to be more credible. Christopher Lovelock lists these sources of information as ways in which customers can reduce the risk of purchasing a service:

- Seeking information from respected personal sources (family, friends, peers).
- Relying on a firm that has a good reputation.
- Looking for guarantees and warranties.
- Visiting scenic facilities or trying aspects of the service before purchasing.
- Asking knowledgeable employees about competing services.
- Examining tangible cues or other physical evidence.
- Using the Internet to compare service offerings.

By gathering information, consumers increase their awareness and knowledge of available choices and product features. A company must design its marketing mix to make prospects aware of and knowledgeable about the features and benefits of its products or brands. If it fails to do this, it has lost the opportunity to sell to the customer. A company must also gather information about competitors and plan a differentiated appeal.

Kotler avers that: .

> Consumers can obtain information from any of several sources. These include personal sources (family, friends, neighbours, acquaintances), commercial sources (advertising, salespeople, dealer and manufacturer, web and mobile sites, packaging, displays), public sources (mass media, consumer rating organization, social media, online searchers and peer reviews) and experimental sources (examining and using the product).
>
> **(Kotler 2017, p. 156)**

Marketers should therefore carefully identify consumers' sources of information and the importance of each source. Consumers should be asked how they first heard about the brand, what information they received, and the importance they place on different information sources. This information helps prepare effective communication.

Evaluation of Alternatives

We have seen how the consumer uses the information on a product to arrive at a set of final brand choices. But how does the consumer choose among the alternative brands? How does the consumer mentally sort and process information to arrive at brand choices? Unfortunately, there is no simple and single evaluation process used by all consumers or even by one consumer in all buying situations. There are several evaluation processes.

Rosemary Martinet preferred a German restaurant with good food and service. However, she believed that all the restaurants under consideration offered these attributes. She also wanted to patronise a restaurant with entertainment and a romantic atmosphere. Finally, she had a limited amount of money, so the price was important. If several restaurants met her criteria, she would choose the one with the most convenient location.

Certain basic concepts will help explain the consumer evaluation processes. First, we assume that each consumer sees a product as a bundle of attributes. For restaurants, these attributes include food quality, menu selection, quality of service, atmosphere, location, and price. Consumers vary as to which of these attributes they consider relevant. The most attention is paid to attributes connected with their needs.

Second, the consumer attaches different degrees of importance to each attribute. That is, each consumer attaches importance to each attribute according to his or her unique needs and wants. Third, the consumer is likely to develop a set of beliefs about where each brand stands on each attribute. The set of beliefs held about a particular brand is known as the brand image. The consumer's beliefs may vary from true attributes because of the consumer's experience and the effects of selective perception, selective distortion, and selective retention. Fourth, the consumer is assumed to have a utility function for each attribute. A utility function shows how the consumer expects total product satisfaction to vary with different levels of different attributes. Fifth, the consumer arrives at attitudes towards the different brands through some evaluation procedure. One or more of several evaluation procedures are used, depending on the consumer and the buying decision.

Purchase Decision

In the evaluation stage, the consumer ranks brands in the choice set and forms purchase intentions. Generally, the consumer will buy the most preferred brand, but two factors can come between the purchase intention and the purchase decision. Attitudes of others represent the first. Rosemary Martinez selected a German restaurant because her boyfriend likes German food. Rosemary's choice depended on the strength of another person's attitude towards her buying decision and on her motivation to comply with those wishes. The more intense the other person's attitude and the closer that person is to the decision-maker, the more influence the other person will have. Nowhere is this better identified than in the case of children. Children do not hide their desires and parents and grandparents are affected intensely.

Purchase intention is also influenced by unexpected situations. The consumer forms a purchase intention based on factors such as expected family income, expected price, and expected benefits from the product. When the consumer is about to act, unexpected situations may arise to change the purchase intention. Rosemary Martinez may have an unexpected car problem that will cost $200 to repair. This may cause her to cancel the dinner reservation and select a less expensive gift.

Because the customer does not know what the experience will be until after the purchase, managers must remember that first-time customers are not really customers, they are only trying the product. While the customer is in the purchase act, employees must do everything possible to ensure they have a good experience and the post-purchase evaluation will be favourable.

Post-purchase Behavior

The marketer's job does not end when the customer buys a product. Following a purchase, the consumer will be satisfied or dissatisfied and will engage in post-purchase actions of significant interest to the marketer. What determines post-purchase satisfaction or dissatisfaction with a purchase? It depends on the relationship between consumer expectations and perceived product performance. If the product matches expectations, the consumer will be satisfied. If it falls short, the consumer will experience a dissatisfaction.

Consumers base their expectations on past experiences and on messages they receive from sellers, friends, and other information sources. If a seller exaggerates the product's likely performance, the consumer will be disappointed. The larger the gap between expectations and performance the greater the consumer's dissatisfaction. This suggests that sellers must faithfully deliver what they promise so that buyers are satisfied. For example, Bermuda enticed tourists to enjoy the island during the off-season at a lower price. They called this period "Rendezvous Time" and advertised that all the island's amenities would be available. When tourists arrived, they found that many facilities and attractions were closed. Hotels had shut down many of their food and beverage facilities, leaving tourists disappointed. Advertising claims initially brought tourists, but the truth got out and hotel occupancy dropped by almost 50 percent for six years.

Summary

In this chapter, we discussed the elements of the stimulus–response model of consumer behaviour. We also outlined the major characteristics affecting consumer behaviour, and examined some of the specific cultural, social, personal, and psychological factors that influence consumers. We finally explained the buyer decision process and discuss need recognition, information search, and evaluation of alternatives, the purchase decision, and post-purchase behaviour.

Activity

1. Examine some of the typical factors affecting the way consumers behave when choosing destination sites.
2. Outline and briefly explain the consumer buying decision process.
3. Briefly explain the personality factors that influence consumer buying behaviour in the hospitality and tourism industry.

References

Kim, J. J., Lee, M. J., & Han, H. (2020). Smart hotels and sustainable consumer behavior: Testing the effect of perceived performance, attitude, and technology readiness on word-of-mouth. *International Journal of Environmental Research and Public Health, 17*(20), 7455.

Kotler, P. (2017). *Principles of marketing*. Seventh European Edition. Pearson Education.

Chapter 9

Organisational Buyer Behaviour

Chapter Outline

- Introduction
- The Organisational Buying Process
- Types of Decisions and the Decision Process
- Participants in the Organisational Buying Process
- Major Influences on Organisational Buyers
- Environmental Factors
- Organisational Factors
- Interpersonal Factors
- Individual Factors
- Organisational Buying Decisions

Objectives

After reading this chapter, you should be able to:

- Explain the organisational buying process.
- Discuss the importance of the participants in the organisational buying process.
- Identify the major influences on organisational buyers.
- List the eight stages of the organisational buying process.

DOI: 10.4324/9781032688497-9

Introduction

Compared with consumer purchases as we examined in the previous chapter, a business purchase usually involves more buyers and a more professional purchasing effort. Also, business purchasing usually involves purchases in large quantities of goods and services to be used as inputs for further outputs.

The Organisational Buying Process

Organisational demand is derived demand; it comes ultimately from the demand for consumer goods or services. It is derived or a function of the businesses that supply the hospitality and travel industry with meetings, special events, and other functions. The American Marketing Association plans and hosts conferences because its members who are marketing managers, suppliers, and educators, have attended past conferences organised by the association. If a particular conference receives poor attendance, the AMA drops it from future schedules. Ultimately, the demand for the AMA members' products determines the demand for AMA products. For example, if a hotel experiences a decline in bookings, management will cut the travel budget of its marketing department. Both events will cause attendance at AMA conferences to fall, resulting in less revenue for the hotels and cities that host these events. Through good environmental scanning, marketers can identify emerging industries, companies, and associations. They screen these organisations to find good business partners.

Compared with consumer purchases, a business purchase usually involves more buyers and a more professional purchasing effort. Corporations that frequently use hotels for meetings may hire their meeting planners. Professional meeting planners receive training in negotiating skills. They belong to associations such as Meeting Planners International, which educates its members on the latest negotiating techniques. A corporate travel agent's job is to find the best airfares, rental car rates, and hotel rates. Therefore, hotels must have well-trained salespeople to deal with well-trained buyers, creating thousands of jobs for salespeople. Additionally, once the meeting is sold the account is turned over to a convention service manager who works with the meeting planner to ensure that the event is produced according to the meeting planner's expectations. Outside the hotel, jobs relating to meetings include corporate meeting and association meeting planners, independent meeting planners, specialty chanellers (intermediaries that usually specialise in incentive travels and meetings) and convention and visitors bureau salespersons.

Types of Decisions and the Decision Process

Organisational buyers usually face more complex buying decisions than individual buyers. Their purchases often involve large sums of money, complex technical features (room sizes, room setups, breakout room, audio-visual equipment, and the like), economic considerations, and interactions among many people at all levels of the organisation. The organisational buying process tends to be more formalised than that of the individual consumer and requires a more professional purchasing effort. The more complex the purchase, the more likely it is that several people will participate in the decision-making process. The total bill for a one-day sales meeting for 20 people can be several thousand dollars. If IBM is to have a series of sales meetings around the country, it will be worthwhile for the company to get quotations from several hotels and spend time analysing the bids.

Finally, in the organisational buying process, buyers and sellers are often very dependent on each other. Sales has become a consultative process. The hotel staff develop interesting and creative menus, theme parties, and coffee breaks. The hotel's convention service staff work with meeting planners to solve their problems. In short, the hotel's staff roll up their sleeves and work closely with their corporate clients to find customised solutions to their needs. When management at the Sands Exposition Centre in Las Vegas discovered that there was insufficient floor space to accommodate a major automobile parts trade show, the Sands rented a temporary 40,000-square-foot pavilion. To help attract attendees to the pavilion, the Sands positioned a restaurant at the back of the facility. In the end, hotels and catering firms retain customers by meeting their current needs and thinking ahead to meet the customer's future needs.

Participants in the Organisational Buying Process

The decision-making unit of a buying organisation sometimes called the buying centre, is defined as "all those individuals and groups who participate in the purchasing decision-making process, who share common goals and the risks arising from the decisions."

The buying centre includes all members of the organisation who play any of the six roles in the purchase decision process:

Users: Users are those who will use the product or service. For instance, a human resource manager who organises a training workshop and elicits the services of a meeting planner. Users often initiate the buying proposal and help define product specifications. If attendees of a sales meeting have a poor experience, they will usually be able to influence the company against using that hotel in the future.

Influencers: Influencers directly influence the buying decision but do not make the final decision. They usually help define product specifications and provide information for evaluating alternatives. For instance, past presidents of trade associations may exert influence on the choice of a meeting venue. Executive secretaries, a spouse, regional managers, and many others can and do exert considerable influence in the selection of sites for meetings, seminars, conferences, and other group gatherings.

Deciders: Deciders select product requirements and suppliers. For example, a company's sales manager for the Denver area will select the hotel and negotiate the arrangements when the regional sales meeting is held in that area.

Approvers: Approvers authorise the proposed actions of deciders or buyers. Although the Denver sales manager arranges the meeting, the contracts may need to be submitted to the corporate vice president of marketing for formal approval.

Buyers: Buyers have formal authority for selecting suppliers and arranging the terms of purchase. Buyers may help shape product specifications and play a major role in selecting vendors and negotiating.

Gatekeepers: Gatekeepers have the power to prevent sellers or information from reaching members of the buying centre. For example, a hotel salesperson calling on a meeting planner may have to go through a secretary. This secretary can easily block the salesperson from seeing the meeting planner. This can be accomplished by failing to forward messages, telling the salesperson the meeting planner is not available, or simply telling the meeting planner not to deal with the salesperson.

Buying centres vary by the number and type of participants. Salespersons calling on organisational customers must determine the following:

- Who are the major participants in the decision-making process?
- What decisions do they influence?
- What is their level of influence?
- What evaluation criteria does each participant use?

When a buying centre includes multiple participants, they may not have the time or resources to reach all of them. Smaller sellers concentrate on reaching the key buying influencers and deciders. It is important not to go over the decider's head. Most deciders like to feel in control of the purchasing decision; going over a decision like working with the boss will be resented. In most cases, the boss will leave the decision up to the decider, and the ill will created by not dealing with the decider directly will result in him/her choosing another company. Larger sellers utilise multilevel, in-depth selling to reach as many buying participants as possible. Their salespeople virtually "live" with their high-volume customers.

Major Influences on Organisational Buyers

Organisational buyers are influenced by several factors when making buying decisions. Some vendors assume that the most important influences are economic. They see buyers as favouring the supplier who offers the lowest price, best product, or most service. This view suggests that hospitality marketers should concentrate on price and cost variables.

Others believe that buyers respond to personal factors such as favours attention, or risk avoidance. A study of buyers in ten large companies concluded that emotions and feelings play a part in the decision process of corporate decision-makers. They respond to "image," buy from known companies, and favour suppliers who show them respect and personal consideration. They "overreact" to real or imagined slights, tending to reject companies that fail to respond or delay in submitting bids.

> The dominant perspective on organizational buying behaviour suggests that buyers tend to rely on objective criteria when making product choice decisions and that the potential influence of subjective cues, such as brands, on buyer decision making decreases with increasing risk.
>
> **(Brown et al., 2011)**

In reality, organisational buyers commonly respond to both economic and personal factors. Where there is substantial similarity in supplier offers, price becomes an important determinant. When competing products differ substantially, buyers are faced with many decision variables other than price comparisons.

The various influences on organisational buyers may be classified into four namely environmental, organisational, interpersonal, and individual.

Environmental Factors

Organisational buyers are heavily influenced by the current and desired economic environment. Factors such as the level of primary demand, the economic outlook, and the cost of money are important. In a recession, companies cut their travel budgets, whereas, in good times, travel budgets are usually increased.

Organisational Factors

Each organisation has specific objectives, policies, procedures, organisational structures, and systems related to buying. It is expected that hospitality marketers become familiar with them as much as possible and will want to know the following: the number of people who are involved in the buying decision, those who take buying decisions, the evaluation criteria, the company's policies, and constraints on buyers.

Interpersonal Factors

The buying centre usually includes several participants, with differing levels of interest, authority, and persuasiveness. Hospitality marketers are unlikely to know the group dynamics that take place during the buying decision process; however, salespeople commonly learn the personalities and interpersonal factors that shape the organisational environment and provide useful insight into group dynamics.

Individual Factors

Each participant in the buying decision process has personal motivations, perceptions, and preferences. The participant's age, income, education, professional identification, personality, and attitudes towards risk all influence the buyer decision-making process. Buyers exhibit different buying styles, therefore hospitality marketers must know their customers and adapt their tactics to known environmental, organisational, interpersonal, and individual influences.

Organisational Buying Decisions

Organisational buyers do not buy goods and services for personal consumption but for further production. They buy hospitality products to provide training, reward employees and distributors, and provide lodging for their employees. There are eight stages of the organisational buying process. This model is called the buy grid framework. The eight steps are as follows:

1. Problem Recognition

The buying process begins when someone in the company recognises a problem or need that can be met by acquiring a good or service. Problem recognition can occur because of internal or external stimuli. Internally, the development of a new product may create the need for a series of meetings to explain the product to the sales force. A human resource manager may notice a need for employee training and set up a training meeting. A CEO may feel that the executive team would benefit from a weekend retreat to reformulate the firm's strategy. With regard to external stimuli, the buyer sees an ad or receives a call from a hotel sales representative who offers a favourable corporate programme. Marketers can stimulate problem recognition by developing ads and appealing to prospects.

2. General Need Description

Having recognised a need, the buyer then determines the requirements of the product and formulates a general need description. For a training meeting, this would include food and beverage, meeting space, audio-visual equipment, coffee breaks, and guestroom requirements. The corporate meeting planner will work with others, including the director of human resources, the training manager, and potential participants to gain insight into the requirements of the meeting. Together, they determine the importance of the price, meeting space, guestrooms, food and beverage, and other factors.

The hotel marketer could assist the buyer during this phase. Often, the buyer is unaware of the benefits of various product features. Marketers can help buyers define their companies' needs and show that their hotel can satisfy them.

3. Product Specification

Once the general requirements have been determined, the specific requirements for the meeting can be developed. For example, a meeting might require 20 guestrooms, a meeting room for 25 set up in classroom style with a whiteboard and overhead projector, and a dining room for lunch. For larger meetings with an exhibition area, the information needed is more comprehensive. These include the availability of water, ceiling heights, door widths, security, and emergency procedures, and other logistics for the event. A salesperson must be prepared to answer their prospective client's questions about their hotel's capabilities to fulfil the product specification.

4. Supplier Search

Based on the product specifications, the buyer conducts a supplier search to identify the most appropriate hotels. The buyer could examine trade directories, do a computer search, or call familiar hotels. The meeting planner will then visit hotels that are found to be suitable, who will develop a short list of qualified suppliers.

5. Proposal Solution

Once the meeting planner has drawn up a short list of suppliers, qualified hotels will be invited to submit proposals. In order to capture the attention and interest of buyers, hotel marketers must be skilled in researching, writing, and presenting proposals. These should be marketing-oriented, not simply technical documents. They should position their company's capabilities and resources so that they stand out from the competition. Many hotels have developed videos for this purpose.

6. Supplier Selection

At this stage, members of the buying centre review the proposals and move towards supplier selection. They conduct an analysis of each hotel, considering their physical facilities, their ability to deliver the service, and the level of professionalism of employees. Frequently, the buying centre specifies desired supplier attributes and suggests their relative importance. The buying centre may attempt to negotiate with preferred suppliers for better prices and terms before making the final selection. There are several ways the hotel marketer can counter the request for a lower price. For example, the dates can be moved from a high-demand period to a need period for the hotel, or menus can be changed. The marketer can add to the value of the services the buyer now receives, especially where services are superior to competitors.

7. Order–Routine Specification

The buyer now makes the final order from the chosen hotel, listing the technical order–routine specifications of the meeting. The hotel will respond by offering the buyer a formal contract. The contract will specify cut-off dates for room blocks, the date when the hotel will release the room block for sale to other guests, payment terms, and minimum guarantees for food and beverage functions. Many hotels and restaurants have turned what should have been a profitable banquet into a loss by not having or enforcing minimum guarantees.

8. Performance Review

The buyer does a post-purchase performance review of the product. During this phase, the buyer will determine if the product meets the buyer's specifications and if the buyer will purchase from the company again. Hotels need to have at least daily meetings with a meeting planner to make sure everything is going well and to correct those things that did not go well. This manages the buyer's perceived service and helps to avoid a negative post-purchase evaluation by the buyer.

Summary

In this chapter, we have discussed the organisational buying process, and identified and discussed the importance of the participants in the organisational buying process. We also examined the major influences on organisational buyers and finally studied the eight stages of the organisational buying process.

Activity

1. Outline and briefly explain the organisational buying decision process.
2. Discuss the importance of each participant in the organisational buying process.
3. How will the managers in the tourism industry in Africa use the organisational buying behaviour to attract organisations to visit tourist sites.

Reference

Brown, B. P., Zablah, A. R., Bellenger, D. N., & Johnston, W. J. (2011). When do B2B brands influence the decision making of organizational buyers? An examination of the relationship between purchase risk and brand sensitivity. *International Journal of Research in Marketing, 28*(3), 194–204.

Chapter 10

Market Segmentation

Chapter Outline

- Introduction
- Markets
- Market Segmentation
- Geographic Segmentation
- Demographic Segmentation
- Psychographic Segmentation
- Behavioural Segmentation
- Requirements for Effective Segmentation

Objectives

By the end of this chapter, you should be able to:

- Explain markets, and market segmentation, and identify several possible bases for segmenting consumer markets, business markets, and international markets.
- Understand the requirements for effective segmentation: measurability, accessibility, substantiality, and actionability.

Introduction

There is the recognition that organisations that sell to individuals and organisations cannot appeal to all buyers in those markets or at least not to all

DOI: 10.4324/9781032688497-10

buyers in the same way. This is because buyers are too numerous, widely scattered, have varied needs, and engage in different buying practices. There is therefore the need to segment the market. This section seeks to explain the forms of segmentation and how to segment the market.

Markets

The term market has acquired many meanings over the years. In its original meaning, a market was a physical place where buyers and sellers gathered to exchange goods and services. To an economist, a market is all the buyers and sellers who engage in the exchange of goods and services. Thus, the fast-food market for instance consists of many sellers, such as Burger King, McDonald's, and Kentucky Fried Chicken, and all the consumers who buy fast-food meals. To a marketer, a market is the set of all actual and potential buyers of a product or service.

Since buyers are numerous, widely dispersed and varied in their needs and engage in different buying practices, organisations cannot appeal to all buyers in the same way. In such a situation, it is better for the organisation to focus on the segment of the market that it can best serve.

Market Segmentation

Buyers tend to differ in their wants, resources, locations, buying attitudes, and buying practice because they have unique needs and wants. Thus each buyer is potentially a separate market. Ideally, a seller might design a separate marketing programme for each buyer. For example, a caterer can customise the menu, entertainment, and setting to meet the needs of a specific client. However, most companies are unable to offer complete segmentation. The cost of complete segmentation is high, and most customers cannot afford completely customised products. Companies, therefore, look for larger groups of buyers who differ in their product needs or buying responses. For example, married adults who vacation with small children have different needs and wants than vacationing young single adults. Thus, Club Meet has developed resorts for families and resorts for couples without children.

Hospitality and tourism businesses have usually employed segmentation in order to properly serve their target markets. Segmentation has enabled these businesses to better understand their customers in terms of their characteristics, tastes, and preferences and thereby adopted the appropriate marketing mix to increase total revenue and business performance. A pioneer of market segmentation in the hotel industry was Quality International (which later became Choice Hotels International Group) in the USA, which in 1981 created different product levels of hotel services for different customer groups. Subsequently, different hotel groups such as Marriot, Hilton, Accor, and Intercontinental Hotel Group (IHG) have continued this practice. Accor for instance has different products for different market segments. These are luxury (Raffles, Fairmont, Sofitel), premium (MGallery, Pullman, Swissôtel), midscale (Novotel, Mercure, Adagio), and economy (ibis, hotelF1) (Khan et al., 2018).

Because each customer group in the food-service market will want a different product, a restaurant cannot reach out to all customers with the same products and services. The restaurant must distinguish the easily accessible consumer groups from those that are hard to reach and the responsive segments from the unresponsive ones. To gain an edge over its competition, a restaurant must examine market segments by identifying one or more subsets of customers within the total market and concentrating efforts on meeting their needs.

Dolnicar (2022), stated that market segmentation is a well-established and commonly used concept in tourism. Businesses and destinations benefit from a segmentation strategy because it allows them to focus on a clearly defined subset of consumers which they are best suited to serve, thus developing a long-term competitive advantage.

There is no single way to segment a market. A marketer has to try different segmentation variables alone and in combination, hoping to find the best way to view the market structure. The following are major variables that might be used in segmenting consumer markets. Here we look at the major geographic, demographic, psychographic, and behaviourist variables used in segmenting consumer markets.

Geographic Segmentation

Geographic segmentation involves dividing the market into different geographic units, such as nations, states, regions, counties, cities, or neighbourhoods. A company decides to operate in one, a few, or all geographic areas paying attention to geographic differences in needs and wants. Fast-food companies often vary their menus to take regional tastes into account. For instance, people in the west want stronger coffee than people in the east. So a coffee shop with branches in London and Beijing will take this into consideration when offering coffee in its shops in the two cities. Also, many tourist destinations tend to segment their tourist markets based on the geographic location of tourists. For instance, Ghana promotes its heritage attractions like the forts and castles linked to the transatlantic slave trade among the African diaspora in Europe and the Americas.

Demographic Segmentation

Demographic segmentation consists of dividing the market into groups based on demographic variables such as age and life cycle, gender, income, occupation, education, religion, race, and nationality. Demographic factors are the most popular basis for segmenting customer groups. One reason is that consumer needs, wants, and preferences often vary closely with demographic variables. Another is that demographic variables are easier to measure than most other types of variables. Even when market segments are first defined using other variables, such as personality or behaviour, demographic characteristics must be known to assess the size of the target market and to reach it efficiently. Hilton for instance targets a customer segment representing middle- and senior-aged professionals with a high level of income who belong to the upper social class.

Consumer needs and wants change with age. Some companies offer different products or marketing strategies to penetrate various age and life-cycle segments. Brochado et al. (2022) concludes that tourists from different demographic segments exhibit heterogeneous patterns regarding perceived value dimensions which influence their behavioural intentions.

Psychographic Segmentation

Psychographic segmentation divides buyers into different groups based on their social class, lifestyle, and personality characteristics. People in the same demographic group can have very different psychographic profiles. Psychographic variables commonly used by tourism marketers are social class, lifestyle, and personality.

Social Class

In Chapter 6 we described the six American social classes and explained that social class has a strong effect on preferences for cars, clothes, home furnishings, leisure activities, reading habits, and retailers. Afternoon tea at the Ritz-Carlton is aimed at the upper-middle and upper-class neighbourhood pub near a factory that targets the working class. The customers of each of these establishments would probably feel uncomfortable in the other establishment.

Lifestyle

Chapter 6 also showed the influence of people's lifestyles on the goods and services that they buy. Marketers are increasingly segmenting their market by consumer lifestyles. For example, nightclubs are designed with certain clientele in mind: young singles who want to meet the opposite sex, singles who want to meet the same sex, and couples who want to avoid singles and enjoy each other's company.

One of the beautiful private rooms at The Sign of the Dove in New York, targets the business community and more affluent customers.

Personality

Marketers also use personality variables to segment markets, endowing their products with personalities who correspond with those of consumers. Persons with similar personality traits are grouped together. Common personality types are introverted, extroverted, emotional, aggressive, etc. For example, Southwest Airlines developed a promotion showing seniors having fun scooting around on snowmobiles. The setting of the ad would have been just as appropriate for 20-year-olds. Southwest was appealing to active seniors, who still viewed themselves as young. The airline was appealing to the kid inside all adults.

Behavioural Segmentation

In behavioural segmentation, buyers are divided into groups based on their knowledge, attitude, use, or response to a product. Many marketers believe that behaviour variables are the best starting point for building market segments.

Occasions

Buyers can be grouped according to occasions when they get the idea, make the purchase, or use a product. Occasion segmentation helps firms to build product use. For example, air travel is triggered by occasions related to business, vacations, or family. Airline advertisements aimed at the business traveller often incorporate service, convenience, and on-time-departure benefits in the offer. On the other hand, airline marketing aimed at the vacation traveller utilises price, interesting destinations, and pre-packaged vacations. Airline marketing aimed at the family market often shows children traveling alone to visit a relative, under the watchful eye of an airline employee. A message of this nature is particularly relevant to the single-parent segment.

Occasion segmentation can help firms to build product use. For example, Mother's Day has been promoted as a time to take your mother or wife out to eat. St. Patrick's Day has been promoted as a night of celebration. Holidays, such as Labour Day and Memorial Day, have been promoted as times to enjoy a mini-vacation. These are examples of occasion marketing.

The honeymoon market represents an occasion with excellent potential for the hospitality industry. In many cultures, the honeymoon trip is paid for by parents or other family members. As a gift, the honeymoon package may contain upscale products and services such as a hotel suite and first-class airfare.

One of the most unusual examples of occasion segmentation is the "Room at the Inn" programme offered by Double Tree Hotels of Canadian Pacific Hotels and Resorts. Double Tree offers free short-term lodging for travellers needing emergency lodging between Thanksgiving and Charismas. These are persons who travel to visit loved ones undergoing emergency medical treatment. Local hospitals, the Red Cross, and United We provide referrals of eligible guests.

Benefits Sought

Buyers can also be grouped according to the product benefits they seek. After studying patrons and non-patrons of three types of restaurants—family popular, atmosphere, and gourmet—one researcher concluded that there are five major appeal categories for restaurant customers. The relative importance of food quality, menu variety, price, atmosphere, and convenience factors across each group. It was found that patrons of family-service restaurants sought convenience and menus. A variety of patrons of atmosphere as restaurants ranked food quality and atmosphere as the top attributes while patrons of gourmet restaurants valued quality.

Knowing the benefits sought by customers is useful in two ways. First, managers can develop products with features that provide the benefits their customers are seeking. Second, managers communicate more effectively with their customers if they know what benefits they seek. For example, some hotel customers like the benefit of feeling refreshed after exercising in a health club. Thus, they will seek out hotels that have health clubs. Other guests do not value exercising. To these guests, the health clubs will not add value to their hotel stay. Thus, a benefit is a positive outcome received from a product feature. Those product features that create positive outcomes for guests create value. Those features that create no positive outcomes for the guest will have no value. By understanding what benefits a customer is seeking marketers can communicate to the guest about features that will create the desired benefits. They can also develop products that will create the benefits customers are seeking.

User Status

Many markets can be segmented into non-users, former users, potential users, first-time users, and regular users. High-market-share companies such as major airlines are particularly interested in keeping regular users and attracting potential users. Potential users and regular users often require different marketing appeals.

Requirements for Effective Segmentation

Although there are many ways to segment a market, all are not equally effective. For example, buyers of restaurant meals could be divided into blond and brunette customers. But hair colour does not affect the purchase of restaurant meals. Furthermore, if all restaurant customers buy the same amount of meals each month, believe all restaurant meals are of equal quality, and are willing to pay the same price, the company would not benefit from segmenting this market.

To be useful, market segments must have the following characteristics:

Measurability: This is the degree to which the segment's size and purchasing power can be measured. Certain segmentation variables are difficult to measure. For instance, the size of the segment of teenagers who drink primarily to rebel against their parents.

Accessibility: This refers to the degree to which segments can be accessed and served. One of the authors found that 20 percent of the customers of a college restaurant were frequent patrons. However, frequent patrons lacked any common features. They included faculty, staff, and students. There was no usage difference among the part-time, full-time, or class years of the students. Although the market segment had been identified there was no way to access the heavy-user segment (Khare & Bhagwate, 2019).

Substantiality: This is the degree to which segments are large or profitable enough to serve as markets. A segment should be the largest possible homogeneous group that is economically feasible to support a tailored marketing programme. For example, large metropolitan areas can support many different ethnic restaurants, but in smaller towns, Thai, Vietnamese, and Moroccan food restaurants would not survive.

Actionability: The degree to which effective programmes can be designed for attracting and serving segments. A small airline, for example, identified seven market segments, but its staff and budget were too small to develop separate marketing programmes for each segment.

Summary

In this chapter, we discussed market segmentation and identified several possible bases for segmenting consumer markets, business markets, and international markets. We then examined the requirements for effective segmentation: measurability, accessibility, substantiality, and actionability.

Activity

1. Define markets in relations to the H&T industry.
2. Explain market segmentation and the five basis for market segmentation.
3. As an executive in the hospitality and tourism sector, explain the factors you consider to effectively segment your market.

References

Brochado, A., Cristovao Verissimo, J. M., & de Oliveira, J. C. L. (2022). Memorable tourism experiences, perceived value dimensions and behavioral intentions: A demographic segmentation approach. *Tourism Review*, *77*(6), 1472–1486.

Dolnicar, S. (2022). Market segmentation for e-tourism. In *Handbook of e-Tourism* (pp. 849–863). Cham: Springer International Publishing.

Khan, Y. H., Hakeem, S. M. A., & Naumov, N. (2018). The use of branding and market segmentation in hotel marketing: A conceptual review. *Journal of Tourism Intelligence and Smartness*, *1*(2), 12–23.

Khare, A., & Bhagwat, J. (2019). The relevance and utility of market segmentation for the international hospitality industry. *Atithya: A Journal of Hospitality*, *5*(1), 4–7.

Chapter 11

Market Targeting and Positioning

Chapter Outline

- Introduction
- Evaluating Market Segments
- Segment Size and Growth
- Segment Structural Attractiveness
- Company Objectives and Resources
- Choosing and Implementing a Positioning Strategy
- Product Differentiation
- Physical Attribute Differentiation
- Service Differentiation
- Personnel Differentiation
- Location Differentiation
- Image Differentiation
- How Many Differences Are necessary for a Positioning Commitment?
- Communicating and Delivering the Chosen Position

Objectives

By the end of this chapter, you should be able to:

- Provide an overview of the process involved in evaluating market segments and propose several techniques for selecting market segments.

DOI: 10.4324/9781032688497-11

- Demonstrate the concept of positioning for competitive advantage by giving specific instances.
- Understand factors influencing choosing and implementation of a positioning strategy.
- Contrast positioning based on product service, personnel, and image differentiation.

Introduction

The final two stages in developing a market-oriented strategy (specifically, market targeting and positioning) are comprehensively examined in this chapter. Market segmentation is a valuable tool for uncovering the vast array of opportunities that exist for a company within its target market. Targeting involves the deliberate choice of the segment(s) that the company intends to cater to. The target market refers to the collection of individuals or organisations whose distinct needs a product is solely meant to fulfil. The company must now thoroughly assess the different segments and determine the number and specific ones to target. We will now delve into the process of how companies evaluate and choose their target markets.

Evaluating Market Segments

When considering various market segments, it is imperative for a company to thoroughly assess three critical elements: the size and potential for growth within the segments, the attractiveness of the segment's structure, and the alignment of the company's objectives and available resources.

Segment Size and Growth

The initial step for the company is to gather and assess data regarding the existing sales of various market segments, their growth rates, and the projected profitability. The company is highly interested in segments that possess dimensions and growth characteristics that are perfectly suitable for their business objectives. However, determining the precise dimensions and growth patterns requires a subjective evaluation. Some companies strategically aim to target segments that exhibit substantial current sales, an exceptionally rapid growth rate, and an impressively considerable profit margin. Nonetheless, it is noteworthy to state that the largest and fastest-increasing segments do not invariably possess attractiveness to every firm. Smaller companies may realise that they don't have the necessary expertise and resources to effectively cater to the larger segments, or they may find that these segments are excessively competitive. In such instances, these companies may opt for smaller and seemingly less attractive segments in an absolute sense, but segments that have the potential to generate significantly higher profitability for their business. According to Nykiel (2007), this approach enables organisations to effectively cater to the identified segments.

Segment Structural Attractiveness

Bodolica and Saleh (2020) assert that the hospitality and tourism industry is currently experiencing a multitude of structural changes. Nevertheless, it is important to comprehend the attractiveness of a particular segment within this sector before investing resources in its pursuit and acquiring customers within the same segment. It is possible for an H&T segment to possess favourable dimensions and growth prospects, yet fail to offer profitable returns. To ensure the long-term desirability of a segment, it is absolutely crucial for an H&T company to thoroughly analyse various significant structural factors. These factors play a pivotal role in shaping the attractiveness of a segment. For instance, a segment may lose its allure if it is already saturated with a plethora of strong and assertive competitors. Moreover, the presence of numerous substitute goods, whether existing or potential, can impose limitations on pricing and hinder the potential profits that can be achieved within a segment. A prime illustration of this phenomenon is the entrance of grocery stores into the home-meal replacement market. As their competitiveness in this domain continues to rise, it is inevitable that it will exert a significant influence on the restaurant market. Lastly, it is essential to acknowledge the significant impact that buyers hold in shaping the desirability of a particular market segment. When buyers possess substantial influence over sellers, they possess the power to demand lower prices, heightened quality, and enhanced services, all while fostering fierce competition among competitors. Regrettably, this can lead to a detrimental effect on the profitability of sellers. Notably, prominent purchasers like an airline establishment situated centrally in Dallas, requiring 50 rooms nightly, possess the capability to engage in negotiations that result in reduced prices.

Finally, a segment can be seen as less attractive if it includes suppliers who hold significant power over pricing dynamics or the overall quality and quantity of purchased goods and services. Suppliers often gain a dominant position when they have a considerable size and concentration, when there are few viable alternatives, or when the supplied product is a crucial input. In certain areas, establishments specialising in providing fresh seafood may be limited by a small number of suppliers. When seafood availability becomes scarce, these establishments face restrictions in their ability to determine the prices they can charge their customers.

Company Objectives and Resources

Even when a segment has the necessary dimensions and shows attractive growth potential, it is crucial for the company to conduct a thorough evaluation of its own objectives and resources in relation to that specific segment. Some segments that may seem appealing at first could be quickly disregarded if they are incompatible with the company's long-term goals. Even though these segments may be attractive, they have the potential to divert the company's focus and resources from its primary objective. Furthermore, they could prove to be an unwise decision from an environmental, political, or social responsibility perspective. To illustrate, in recent times, several hotel chains have deliberately chosen not to enter the gaming industry. Even when a segment has the necessary dimensions and shows attractive growth potential, it is crucial for the company to conduct a thorough evaluation of its own objectives and resources in relation to that specific segment. Some segments that might appear attractive initially could be swiftly dismissed if they are inconsistent with the organisation's long-term objectives. Although these segments may be enticing, they could potentially lead the company astray and deplete its resources from its primary objective. Furthermore, they could prove to be an unwise decision from an environmental, political, or social responsibility perspective. To illustrate, in recent times, several hotel chains have deliberately chosen not to enter the gaming industry.

Choosing and Implementing a Positioning Strategy

Positioning requires strategically placing a product or service in the minds of consumers, leveraging its key attributes (Kotler & Amstrong, 2018). This vital process encompasses three essential steps: recognising a multitude of potential competitive advantages to establish a favourable position, meticulously selecting the most suitable advantages, and skillfully conveying and delivering the chosen position to a specifically targeted market segment. This marketing technique holds great significance for tourist brands, particularly for lesser-known destinations seeking entry into the fiercely competitive tourist market (Maralbayeva et al., 2021).

A company possesses the remarkable capability to set itself apart from its rivals by artfully combining its distinctive strengths. This advantageous position is achieved by presenting consumers with the irresistible allure of lower prices for comparable products in contrast to its competitors, or by providing supplementary advantages that unequivocally justify higher prices. Consequently, it is imperative for a company to consistently assess its pricing and product offerings in relation to those of its competitors, while simultaneously exploring potential avenues for enhancement. In the event that the company is able to surpass its competitors, it can be deemed to have successfully attained a competitive advantage.

In specific circumstances, the technique of dividing products has also proven to be an effective positioning strategy. In the era before the early 1970s, numerous destination resorts exclusively offered a comprehensive package known as the American plan (AP), which encompassed a wide array of services including food and beverage. However, in light of evolving consumer preferences, many guests no longer yearned for the rigidity of three meals per day and an obligatory Friday evening formal dance. Recognising this shift in guest behaviour, astute resort managers began to set their properties apart by offering a modified American plan (MAP), where lunch was excluded, or alternatively, a European plan that gracefully omitted any meal offerings.

Not every organisation is fortunate enough to discover a plethora of chances to obtain a competitive advantage. Some companies may only discover minor benefits, which are often easily replicated and therefore short-lived. These companies must consistently search for new potential advantages and gradually introduce them one by one in order to disrupt their competitors. Very few companies, if any at all, are capable of achieving a significant and long-lasting advantage; instead, they obtain smaller advantages that help them expand their market share over time. Hotels, resorts, and restaurants occasionally hold the belief that their beachfront location, proximity to an airport, adjacency to a ski resort, or position in the central business district bestows upon them a permanent advantage. However, historical evidence presents a different reality. Beach erosion and pollution are detrimental to the serene beauty that once attracted countless visitors. Likewise, the decline in popularity of ski resorts is a stark reminder of the fleeting nature of trends. Airports, once bustling hubs of activity, are forced to relocate and adapt to changing circumstances. Even central business districts, once vibrant and alluring, can lose their charm over time. However, it is disheartening to witness the management of hospitality establishments, which possess inherent advantages, display a diminishing concern for their valued customers and dedicated staff. Regrettably, this lack of attentiveness only serves to hasten their inevitable decline.

Product Differentiation

To excel in today's fierce market, a company should aim to separate its product from the offerings of its challengers. It is crucial for a company to find ways to set itself apart and stand out amongst the majority of products out there. By giving priority to the physical features, service, personnel, location, or perception of the brand, a company can effectively separate its offering and attain a competitive advantage over its rivals. The talent to separate is fundamental to excel in the entrepreneurial landscape.

Physical Attribute Differentiation

Hotels such as the Sheraton Place in San Francisco, the Palmer House in Chicago, the Waldorf-Astoria in New York, and the Raffles in Singapore are unquestionably different from the grandeur of the past. Their physical surroundings possess an unmatched allure that a newly constructed hotel simply cannot rival. The atmosphere created by Planet Hollywood, adorned with its remarkable collection of memorabilia from the film industry, and the Hard Rock Cafe, proudly showcasing its wide array of music memorabilia, will undoubtedly be incredibly difficult for rivals to replicate. However, it is important to note that MGM airlines offered an exclusive aircraft meticulously designed to cater solely to first-class passengers. This noteworthy aircraft featured a standing bar, luxurious couches, and other distinctive physical features that were noticeably absent in the first-class compartments of major domestic carriers.

Service Differentiation

Some businesses set themselves apart by providing an exceptional level of service. For example, Sheraton surpasses expectations by providing a trouble-free check-in service directly in the convenience of your own room. Likewise, Red Lobster allows their valued customers to conveniently add their names to a virtual waiting list from the convenience of their own homes, resulting in reduced wait times at the establishment. Certain dining establishments occasionally go the extra mile by delivering tasty meals directly to customers' doorsteps, thereby distinguishing themselves from their competitors. By adapting their services to fulfill the particular requirements of their target audience, companies can acquire a distinct competitive edge in the market.

Personnel Differentiation

Companies can attain a considerable competitive edge by hiring and nurturing exceptional staff, surpassing their rivals. Accordingly, Singapore Airlines has acquired a distinguished reputation primarily because of the elegance shown by its flight attendants. Herb Kelleher, the founder of Southwest Airlines, asserts that it may be possible for a competitor to imitate their low-cost system, but it would be impossible for that competitor to replicate the unique spirit embodied by Southwest's employees.

The process of personnel differentiation necessitates the careful selection and thorough training of customer-facing staff by a company. These individuals must possess the necessary competence, skills, and knowledge. Moreover, they should demonstrate traits of politeness, friendliness, and respectfulness towards customers. It is essential that they consistently and accurately cater to the needs of customers, while also making a concerted effort to comprehend their requirements, communicate effectively, and promptly address any customer inquiries or issues.

Location Differentiation

The strategic location of a business in the realm of hospitality and travel can provide a substantial edge over competitors. Take, for example, the hotels that grace the presence of Central Park in New York City. These magnificent establishments possess a distinct advantage over their counterparts only a block away, devoid of the breath-taking park view. Similarly, motels conveniently situated right off a freeway exit relish in a considerable advantage when it comes to occupancy rates, leaving their counterparts a block away in the dust. Restaurants that perch atop majestic mountains proudly tout their scenic views as a competitive advantage, just as those with a view of the vast ocean do. Not to be outdone, international airlines skillfully leverage their geographical location to set themselves apart in their respective domestic markets. Qantas, for instance, artfully positions itself as Australia's esteemed national carrier, amassing a devoted following in its homeland. Given these compelling examples, it is imperative for hospitality and travel companies to meticulously identify the perks bestowed upon them by their location and employ these invaluable advantages to differentiate themselves in the fiercely competitive market.

Image Differentiation

Even when presented with similar competing offers, buyers may perceive a difference based on the reputation and brand identity of a company. Therefore, it is imperative for companies to strategically establish distinct brand images that set them apart from their competitors. An effective company or brand image should effectively convey a singular and unique message that effectively communicates the significant benefits and positioning of the product. The development of a robust and distinctive image necessitates a combination of creativity and diligent effort. It is not feasible for a company to instantaneously implant an image in the minds of the public through a mere handful of advertisements. Chills, for instance, has successfully cultivated an image as a casual and enjoyable neighbourhood restaurant, effectively communicated through their advertising, menu, physical ambiance, and staff. It is crucial for the image to be consistently supported by every aspect of the company's communication and actions. Studebakers, on the other hand, strategically positioned itself as a singles nightclub catering to adults over the age of 25 years, aligning itself with a specific product class and user demographic. The name of the establishment, derived from a car, further reinforces this positioning. The utilisation of the Studebaker brand and the inclusion of a Studebaker automobile within the establishment did not carry much weight with a younger demographic, who probably had not come across this vehicle on the streets and highways. As a result, the term Studebaker had meaning for their target market, but had little value for younger consumers.

If a company is fortunate enough to discover several potential competitive advantages, it is crucial to carefully choose the advantages that will form the foundation of the positioning strategy. It is of utmost importance to determine the number of distinctions to highlight and which ones to prioritise.

How Many Differences Are Necessary for a Positioning Commitment?

From the perspective of numerous marketers, it is absolutely imperative for companies to vigorously endorse a singular advantage to the target market. For instance, it has been stated by the highly regarded advertising executive, Rosser Reeves, that it is imperative for a company to establish a distinctive selling point (USP) for each brand and steadfastly adhere to it. Each brand should meticulously select a specific attribute and proudly proclaim itself as the foremost in that particular aspect. In an era saturated with communication, individuals have a tendency to retain a stronger memory of the top-ranking option. As a result, Motel 6 consistently emphasises its status as the most budget-friendly national chain, while Ritz-Carlton positions itself as a frontrunner in terms of value. What are some of the paramount positions to prioritise in promotion? The primary ones encompass unparalleled quality, extraordinary service, unmatched affordability, optimal value, and unrivaled location. A company that passionately champions a stance that carries great importance for its desired clientele and consistently fulfills that commitment is bound to emerge as the most renowned and unforgettable.

Other marketers firmly believe that it is imperative for companies to establish their presence based on multiple distinguishing factors. A restaurant, for instance, may boldly assert that it not only serves the most delectable steaks but also provides exceptional service. Similarly, a hotel may boldly declare that it provides unparalleled value and an unbeatable location. Currently, in an era where the mass market is consistently dividing into numerous smaller market segments, companies are tirelessly working to extend their positioning strategies to attract a broader range of segments. A prime example of this can be observed in the Boulders, located in Arizona, which astutely positions itself as a premier golf resort while simultaneously embracing the identity of a luxurious retreat, allowing its esteemed guests to immerse themselves in the breathtaking flora and fauna of the Sonora Desert. By adopting such an approach, the Boulders successfully manages to allure both avid golfers and non-golfers alike.

However, as companies strive to enhance the credibility of their brands, they face the potential risk of scepticism and the detrimental consequences of losing their distinct positioning. In order to ensure success, it is crucial for a company to steer clear of three major positioning pitfalls. The first of these is under-positioning, a grave error that involves neglecting to establish a clear and compelling company identity. Some companies find themselves in the unfortunate situation where potential buyers possess only a vague notion of their existence or are unaware of any noteworthy attributes associated with their brand. This is particularly prevalent among independent hotels attempting to tap into the international market. The Seoul Plaza Hotel, a luxurious establishment located in the vibrant city of Seoul, is a prime example of an under-positioned entity. The remarkable existence of this phenomenon is still shrouded in obscurity across the continents of Europe and North America. To establish a strong and unshakeable presence in the hospitality industry, esteemed hotels like the Seoul Plaza have cleverly taken the strategic position of aligning themselves with esteemed marketing conglomerates such as the prestigious "Leading Hotels of the World" and the highly regarded "Preferred Hotels."

One should however, exercise vigilance against falling victim to the peril of over-positioning, wherein the company inadvertently imparts a confined and restricted perception to potential purchasers. Companies must absolutely avoid any form of confused positioning, as it inevitably leaves buyers with a muddled and unclear perception of the company. To make this point clear, let us examine the case of Burger King, which has faced numerous challenges in establishing a profitable and consistent position throughout the years. Since 1986, Burger King has successfully launched a total of eight distinct advertising campaigns, each one meticulously crafted with its own unique theme and compelling messaging. These campaigns have varied from the memorable "Herb the nerd doesn't eat here" to the captivating "This is a Burger King own," and from the enticing "The right food for the right times" to the rebellious "Sometimes you've gotta break the rules." Remarkably, their most recent successful campaigns have been "Get Your Burger's Worth," followed by the irresistible "It Just Tastes Better" in 1998, and most recently, the captivating "Got the Urge." Despite these achievements, the president of Burger King faced a crucial decision in late 2001. He openly confessed that their advertising agency, which they had relied on for the past eight years, had not been able to create an advertising campaign that truly resonated with customers as powerfully as they desired. This lack of resonance has unfortunately led to consumer confusion and a decline in sales for Burger King franchises.

Which Differences

Not all brand distinctions hold significant value or are worth pursuing. Not every variation can effectively set a brand apart. Each distinction carries the potential to incur costs for the company while also providing benefits to customers. Hence, it becomes imperative for the company to meticulously choose how it will differentiate itself from competitors. The differentiation could be described as valuable if the following criterion are met.

- *Important*: The differentiation is important if it provides value for target customers.
- *Distinctive*: The differentiation is unique such that competitors do not provide the same level of uniqueness, whereas our company has the capability to offer it in a more pronounced manner.
- *Superior*: The distinction is of a higher nature than alternative methods through which customers can obtain the same benefit.
- *Communicable*: The distinction lies in its communicability and its discernibility to potential purchasers.
- *Pre-emptive*: The differentiation is valuable if it becomes very difficult for competitors to copy the difference.
- *Affordable*: Though distinct and provides value for customers, it should be affordable for them.
- *Profitable*: The company can introduce the difference profitably.

Many companies have failed to meet one or more of these tests with their introduced differentiations. The Westin Stamford Hotel in Singapore proudly claims to be the tallest hotel in the world, but this distinction may not matter to many tourists; in fact, it might even discourage some.

Certain competitive advantages may be quickly eliminated due to their slight nature, high development costs, or inconsistency with the company's profile. Let's suppose that a company is in the process of designing its positioning strategy and has narrowed down its list of potential competitive advantages to four. In such a scenario, the company requires a framework to assist in selecting the most sensible advantage to develop.

Communicating and Delivering the Chosen Position

Once an appropriate positioning characteristic has been selected and a positioning statement has been formulated, it becomes imperative for the company to effectively convey its position to the intended customers. The company's entire marketing mix endeavours should align with and bolster its positioning strategy. Should the decision be made to enhance service superiority, for instance, it becomes necessary to recruit service-oriented personnel, implement comprehensive training programmes, incentivise employees for delivering exceptional service, and craft persuasive sales and advertising messages to disseminate the message of superior service.

Developing and upholding a steadfast positioning strategy can be quite challenging. There are numerous opposing forces that consistently come into play. The advertising agencies hired by the company might not value the chosen position and could potentially work against it, either openly or secretly. Furthermore, the new management may struggle to understand the positioning strategy. Additionally, there could be a decrease in budgets for important support programmes like employee training or sales promotion. Establishing an effective position requires a steadfast, long-term plan with unwavering support from management, employees, and vendors.

Companies create memorable statements to clearly convey their desired market position. For example, Burger King's slogan, "Have it your way," effectively communicates to customers that they have the freedom to customise their condiments according to their preferences. Similarly, La Quina's slogan, "Just right overnight," appeals to car travellers who are in need of overnight accommodations without the full-service hotel experience.

Avis Auto Rental initially established its position by expressing a compelling statement and implementing a strong support programme that aimed to convince customers with the message, "We go the extra mile as the second-ranked provider." This positioning strategy also aligned Avis with the leading company, Hertz, while distinguishing it from competitors such as Budget, Dollar, National, and Thrifty. The objective of these statements is to cultivate a positive perception among the intended customer base.

Summary

In this chapter, we have provided an outline of the process involved in evaluating market segments, and we have also proposed several methods for selecting market segments. Moreover, we have taken the opportunity to demonstrate the concept of positioning for competitive advantage by providing specific examples. Subsequently, we have engaged in a detailed discussion regarding the process of selecting and implementing a positioning strategy, and compared various approaches to positioning, including those based on product service, personnel, and image differentiation.

Activity

1. Explain market targeting and the factors that must be considered when selecting a segment(s) to target.
2. What is positioning? Discuss the various positioning strategies available to an H&T institution looking to expand into several African countries.
3. Explain the concept of differentiation and how managers in the H&T industry differentiate their product offerings from that of competitors.

References

Bodolica, V., Spraggon, M., & Saleh, N. (2020). Innovative leadership in leisure and entertainment industry: The case of the UAE as a global tourism hub. *International Journal of Islamic and Middle Eastern Finance and Management*, 13(2), 323–337.

Kotler, P., & Amstrong, G. (2018). *Principles of marketing*. (17th Ed.) . Pearson Education Limited.

Maralbayeva, S. M., Nikiforova, N. V., & Smykova, M. R. (2021). The destination life cycle concept in developing a tourist brand. Case of Mangystau of Kazakhstan. *Journal of Environmental Management & Tourism, 12*(6 (54)), 1472–1494.

Nykiel, R. A. (2007). *Handbook of marketing research methodologies for hospitality and tourism*. Routledge.

Chapter 12

Designing and Building Brands

Chapter Outline

- Introduction
- Product
- Product Levels
- Core Products
- Facilitating Products
- Supporting Products
- Augmented Products
- Branding
- Conditions Supporting Branding
- Easy Identification of Product by Brand or Trademark
- The Product Is Perceived as the Best Value for the Price
- Quality and Standards Are Easy to Maintain
- Leveraging Brand Equity

Objectives

After reading this Chapter, you should be able to:

- Explain the concept of a product, including the core, enabling, supplementary, and enhanced elements of the product.
- Describe the factors to be considered when designing a product, which includes the ambiance, customer engagement with the service delivery

DOI: 10.4324/9781032688497-12

system, customer interaction among themselves, and customer involvement in the production process.
- Elaborate on the concept of branding and the prerequisites that contribute to effective branding.

Introduction

In this chapter, we are going to study how to design and build brands in the tourism and hospitality industry. We shall be looking at the concept of a product and how branding can be applied in marketing tourism and hospitality products.

What Is a Product?

A luxurious accommodation at the esteemed Four Seasons in Toronto, an exotic and breathtaking retreat in the mesmerising Hawaiian islands, the world-renowned and irresistibly delicious McDonald's French fries, an all-inclusive and extraordinary vacation package in the enchanting paradise of Bali, a meticulously prepared and exceptionally catered luncheon, an enlightening and captivating bus tour exploring the rich history of historic sites, and a truly remarkable convention held in a state-of-the-art modern convention centre with exclusive group rates in a conveniently located nearby hotel are all remarkable products that deserve your utmost consideration. Let us now contemplate the sheer diversity and unparalleled range of products available in a typical casino hotel. Indeed, the tourist product itself constitutes a remarkable bundle, artfully combining elements of luxurious accommodations, exquisite dining experiences, seamless transportation services, enthralling entertainment offerings, and an array of other indispensable ancillary services, all harmoniously working together to deliver an unparalleled and unforgettable travel experience.

We define the term product as any offering that holds the power to capture the attention, acquire, and fulfill the desires or needs of a market (Kotler, 2012). This encompasses tangible goods, services, destinations, establishments, and even concepts. It is significant to highlight that this definition specifically relates to the deliberate element of the product that the company showcases. However, alongside this planned aspect, there exists an unplanned component as well. This holds particularly true for products in the hospitality and travel industry, where variations are common. For instance, imagine a scenario where a patron enters a diner in Dallas and is warmly greeted by the hostess, who presents them with a menu. As they open the menu, they are met with an unpleasant surprise – a deceased cockroach clinging to the pages. Naturally, receiving such an unexpected addition, the customer promptly decides to depart from the establishment. Clearly, the presence of a dead roach on the menu was not part of the restaurant's intentions. It is clear that the final result received by the customer may not always align with the meticulous plans of management. Managers of service organisations must exert significant effort in order to eradicate any unforeseen circumstances and guarantee that the guests receive precisely what they have anticipated.

Product Levels

Hospitality managers must carefully ponder over the merchandise from four distinct standpoints: the core merchandise, the facilitating merchandise, the supporting merchandise, and the augmented merchandise.

Core Products

The core product, at its most fundamental level, addresses a crucial question: What is the true essence of what the buyer seeks? Each and every product is essentially a bundle of problem-solving solutions. As Theodore Levin astutely observed, buyers are not in fact purchasing quarter-inch drills; rather, they are seeking quarter-inch holes. In a similar manner, when a guest stays at a hotel, their intention is not to acquire the bed or the food, but rather to locate a haven for repose and rejuvenation. Just as all exceptional steak houses have come to understand, the key lies in not simply selling the steak, but in captivating the customer with the sizzling experience. Marketers must skillfully uncover the core advantage that a product offers to the consumer, and focus on selling these benefits rather than merely promoting its features.

Facilitating Products

Facilitating products are essential for the guests to fully enjoy and utilise the core product. In order to provide an exceptional experience, a top-notch corporate hotel must offer impeccable check-in and checkout services, convenient telephones, a delightful restaurant, and even valet service. On the other hand, in a limited-service economy hotel, the facilitating services may be more modest with just the necessary check-in and checkout service and public phones on the property. Accessibility emerges as a crucial factor when it comes to facilitating products. While resort condominiums may decide to close the office and registration desk in the evenings, they can easily overcome this by informing the guests in advance and making suitable arrangements for late arrivals to collect their keys. Nevertheless, for a business hotel, it is simply inconceivable to close the front desk. The guests have the utmost expectations that it will always remain accessible. Similarly, if a hotel boasts a business centre, it must always ensure its accessibility and availability to cater to the needs of the guests whenever they desire. To produce an undeniably extraordinary product, one must possess an immensely profound understanding of the specific market it is intended for, as well as an extremely sharp awareness of the services that are absolutely necessary for its success.

Supporting Products

Core products necessitate enabling products, while they do not necessitate auxiliary products. Auxiliary products are supplementary offerings that enhance the value of the core product and aid in setting it apart from competitors. In the realm of corporate hospitality, the provision of a business centre or a comprehensive health spa are auxiliary products that may entice customers to choose a particular hotel. The distinction between enabling and auxiliary products is not always straightforward. For instance, products that enable one market segment may serve as auxiliary products for another. Take, for example, the case of families staying at a hotel who may not require access to restaurants or valet service, whereas these amenities are indispensable for business travellers. Hyatt was a frontrunner in offering an extensive range of bathroom amenities, comprising shampoo, conditioners, and a variety of soap options. Upon introducing these amenities, they served as auxiliary products to enhance the core product of hotel rooms. Today, in esteemed establishments like Hyatt and comparable hotels, amenities have transformed into indispensable commodities. Other lodging options promptly followed suit in emulating Hyatt's comprehensive amenity offerings, leading to a growing anticipation among travellers for such services in this particular category of accommodations.

Augmented Products

The enhanced offering encompasses various crucial aspects such as accessibility, ambiance, the interaction customers have with the service provider, customer involvement, and the interaction between customers themselves. These components seamlessly merge with the central, facilitating, and supporting products to create an extraordinary offering.

From a managerial perspective, the central product serves as the focal point for the business, representing its very purpose. Facilitating products are indispensable in delivering the central product to the intended market. Supporting products play a vital role in positioning a product. According to esteemed services marketing expert Christian Gronroos, the central, facilitating, and supporting products determine what the customer receives, although not precisely how they receive it. The method by which the service is provided greatly affects how customers perceive it. The augmented service offering blends what is provided with the manner in which it is delivered.

The concept of the augmented product holds immense significance as it pertains to the realm of hospitality and travel services, where the active participation of customers is required in the creation of the service. In the case of most products, it is the customer who approaches the service delivery system, necessitating their interaction with the said system. This interaction entails various tasks such as checking in at the front desk, navigating their way to the designated room, and comprehending the functionality of the television and phone systems. Furthermore, customers are also expected to engage with fellow patrons and employees. Given that customers physically visit the service, the atmosphere assumes a pivotal role in shaping the overall product. Therefore, the augmented product encompasses the fundamental aspects that demand meticulous management when customers patronise the service facility, be it a hotel, restaurant, country club, conference centre, amusement park, and so forth.

The core product necessitates heightened customer engagement, as well as an enhanced augmented product, in order to optimally situate the product destinations within the hospitality industry.

Branding

Why Brand?

A brand encompasses a name, a term, a sign, a symbol, a design, or a fusion of these elements, all of which serve the purpose of identifying the goods or services offered by a seller and setting them apart from those provided by competitors. Within a brand, the brand name holds the power to be spoken aloud, as exemplified by renowned names like Golden Tulip, Fiesta Royale, McDonalds, Sizzler, Emirates, and Disneyworld. However, a brand mark is a component of a brand that can be visually recognised, yet cannot be articulated, taking the form of a symbol, design, or distinctive coloring or lettering. Examples of this include the iconic golden arches of McDonald's and the distinguished H of Hilton. Furthermore, a trademark is a brand or a segment of a brand that is granted legal protection, ensuring that the seller possesses exclusive rights to utilise the brand name or brand mark.

Brands hold immense value in a company's repertoire, and astute businesses of today are well aware of the significance of leveraging their brands. These enterprises understand that brands encompass much more than mere products and services; they encapsulate the very essence of the company itself. A brand serves as a vital pillar of a company's identity, embodying trust, reliability, and a clear set of expectations. The most influential brands in the world occupy a permanent space in the minds of customers, such that their mere mention evokes a unanimous association.

Conditions Supporting Branding

The following five conditions contribute to the branding decision:

- The product is easy to identify by brand or trademark.
- The product is perceived as the best value for the price.
- Quality and standards are easy to maintain.
- The demand for the general product class is large enough to support a regional, national, or international chain. Developing a critical mass to support advertising and administrative overhead is important.
- There are economies of scale.

Easy Identification of Product by Brand or Trademark

Hotel and restaurant chains offer a plethora of prime examples that effortlessly catch the eye. The eye-catching red-and-white awnings and the distinctive painting of T.G.I. Friday's and Holiday Inn's iconic green signs are instantly recognisable to patrons. A majority of freeway billboards serve as directional signs, relying on the power of brand recognition. They simply showcase the brand name and/or the brand symbol along with directions to the establishment.

The development of a brand name is pivotal in establishing the exclusive identity of the brand. A brand name should possess a range of desirable qualities, including:

- It should suggest something about the product's benefits and qualities. Examples: Daily Queen, Comfort Inns, Pizza Hut, Burger King, Sizzler, American Airlines.
- It should be easy to pronounce, recognise, and remember. Short names help. Examples: Wendy's, Hilton, KLM, KFC, the Shuttle (United's limited-service airline).
- It should be distinctive. Examples: El Torito, Avis, Bennian's.
- For larger firms looking at future expansion into foreign markets, the name should translate easily into foreign languages. Some firms have found that their names have a negative meaning when translated into the language of the countries into which they want to expand. For instance, a UK food manufacturer Sharwoods launched a new Indian "Bundh" sauce range. However, in the Punjabi language, "bundh" sounded just like the Punjabi word for "ass."
- It should be capable of registration and legal protection.

Sometimes, a company reaches a point where it surpasses its original name and must embark on a transformative journey. Notable examples of major chains that have undergone name changes include Western International, which evolved into the sophisticated and refined Westin, and Hilton Hotels, which gracefully transitioned to become Hilton Hotels and Resorts. It is not uncommon for companies to seek to enhance their image by revamping their logos, all while keeping their beloved and respected name intact. However, once a name has been carefully selected, it becomes imperative to safeguard it with unwavering dedication. A case in point is Quality International, now known as Choice Hotels International, which initially opted for the name McSleep for its line of budget hotels. Unfortunately, this choice sparked a legal dispute with McDonald's, leading to the ultimate rebranding of the hotels as Sleep Inns. It is manifest that companies are extremely aware of the significance of safeguarding their trade names, as failure to do so might result in relinquishing exclusive rights and privileges. When it comes to luxury experiences, discerning customers seek a harmonious blend of captivating aesthetic allure and the embodiment of impeccable employee attitudes, both of which significantly shape the overall experience and perception of the brand (Venkatesh et al., 2010; Ward & Dahl, 2014).

The Product Is Perceived as the Best Value for the Price

A brand name holds immense value as it is directly linked to the perception of consumers. By fabricating a perception of top quality and worth, brands are capable of attracting and retaining a faithful consumer base. Take for example La Quinta, which has successfully cultivated a positive image among overnight business travellers. Furthermore, Embassy Suites has positioned itself as a brand that offers extraordinary value for those seeking an all-suites hotel experience. The authority of a brand name is not bounded by products and services alone, but reaches to tourist destinations as well. Think of renowned places like Vail, Aspen, Acapulco, Palm Springs, and the French Riviera, which have all built strong reputations, shaping consumer perceptions and setting high expectations. It is the duty of those involved in promoting and developing tourist destinations to take on the responsibility of enhancing and safeguarding these favourable brand images.

Quality and Standards Are Easy to Maintain

To achieve success, it is imperative for a renowned multiunit brand like Pizza Hut, Holiday Inn, Chili's, or Nathan's Hot Dogs to establish comprehensive and consistent standards that surpass customer expectations. By cultivating an exceptional image of quality, customers inevitably demand excellence from every establishment affiliated with the brand. In order to preserve the brand's value, it is vital to avoid any deviations from the set standards and policies, as consistency and standardisation are the cornerstones of triumph.

It is worth noting that consumers frequently exhibit unwavering loyalty towards specific brands. The real power of branding lies in nurturing a feeling of allegiance among customers, leading to their steadfast dedication to purchase the brand whenever it is made accessible. Remarkably, the more widely accessible the brand becomes, the more influential its name becomes. This is precisely why the majority of renowned hotel chains strive to establish their presence in prominent locations within their respective markets. To achieve success, it is imperative for a renowned multiunit brand like Pizza Hut, Holiday Inn, Chili's, or Nathan's Hot Dogs to establish comprehensive and consistent standards that surpass customer expectations. By cultivating an exceptional image of quality, customers inevitably demand excellence from every establishment affiliated with the brand. In order to preserve the brand's value, it is vital to avoid any deviations from the set standards and policies, as consistency and standardisation are the cornerstones of triumph.

It is worth noting that consumers frequently exhibit unwavering loyalty towards specific brands. The real power of branding lies in nurturing a feeling of allegiance among customers, leading to their steadfast dedication to purchase the brand whenever it is made accessible. Remarkably, the more widely accessible the brand becomes, the more influential its name becomes. This is precisely why the majority of renowned hotel chains strive to establish their presence in prominent locations within their respective markets.

Not all brands can boast the same level of success as McDonald's. According to Peter Yesawich, the esteemed president of Yesawich, Pepperdine, & Brown, the key to a brand's triumph lies in its ability to establish a distinct and unmistakable presence in the minds of its customers. Yesawich asserts that effective advertising must convey the perception of an innovative and groundbreaking product. In order to captivate the customer, a new brand must effectively communicate the array of benefits it offers. Robert Hazzard, the accomplished CEO of Choice Hotels International, asserts that "people are drawn to exceptional bargains. The challenge lies in informing them about the true value of such deals." Hazzard argues that many hotel brands falter in setting themselves apart from their competition by neglecting to highlight why their offerings are superior. He emphasises that while Holiday Inns featured the spectacle of Bugs Bunny leaping across a swimming pool in their advertisements, other major chains failed to convey the tangible benefits that consumers could enjoy. In our advertisements, we proudly featured the captivating Vanna White, esteemed hostess of the renowned television show Wheel of Fortune. In this captivating visual, she gracefully emerged from a suitcase while conveying a compelling message: "Choose to stay with us, and you will not only experience the utmost comfort in our rooms, but you will also be rewarded with a generous offering of one thousand dollars in complimentary discount coupons."

There Are Economies of Scale

Branding necessitates financial investment. It is imperative for the company endorsing a brand to establish and implement robust standards, systems, and quality assurance programmes. The brand name must be actively promoted. In order to rationalise the expenses incurred for administration and advertising, the brand should offer substantial economies of scale. These economies of scale typically encompass reduced promotional costs, as all brand units within the sphere of influence of the advertising campaign reap the benefits of such promotion. The provision of management information systems, reservation systems, national purchasing contracts, and uniform architectural designs are some of the ways in which brands can deliver economies of scale. Quincy's Steak House and Red Roof Inns seem to adopt a strategy of swiftly developing multiple units within a concentrated area. The number of units in a given area serves as a strategic ploy, as it enables the public to suddenly witness a cluster of offerings. Consequently, word-of-mouth promotion ensues as a direct outcome. The allocation of advertising expenses in local or regional media, including newspapers, television, and radio broadcasts, can be strategically distributed across various units. A solitary, independent restaurant, or hotel may not possess the extensive reach of a multi-unit establishment, and therefore lacks the necessary advertising funds to create a substantial impact within a regional or national market.

Leveraging Brand Equity

Brand equity can be effectively harnessed and capitalised upon through the powerful tools of cobranding and strategic partnerships. In certain instances, companies with shared ownership seize the opportunity to present their esteemed brands within a single establishment, harmoniously coexisting under one roof. An exemplary illustration of this is Tricon Global, an extraordinary conglomerate that proudly boasts ownership of the iconic Taco Bell, KFC, and Pizza Hut, all of which are elegantly showcased together. Moreover, Tricon's astute methodology involves housing these distinguished brands in separate units that conveniently reside alongside each other, thereby ingeniously fostering a captivating cross-promotion that effortlessly satiates the cravings of both the eager youngsters yearning for delectable chicken and the discerning mothers seeking the tantalising flavours of Mexican cuisine. Similarly, the beloved Tim Horton's donut shops and Wendy's, both under the umbrella of shared ownership, artfully unite to form a harmonious culinary haven, ubiquitously gracing the Canadian landscape.

Sometimes, companies will opt to join forces with another brand to create a powerful cobranding strategy. Take, for example, Carl's Jrs., who successfully introduced the Green Burrito, a Mexican fast-food restaurant nestled within their already established hamburger joints. Similarly, Blimpie took a bold step by launching Pasta Central, a venture that complements their sandwich shops. Yet, they didn't conclude there. Blimpie also sought out a partnership with Winner's Chicken and Biscuits, an existing fast-food company, to further enhance their cobranding efforts.

Aside from the undeniable appeal of having two renowned brands under one roof, cobranding also brings forth the advantage of efficiency and scalability. Joseph Morgan, the esteemed group president of Blimpier, rightly emphasises that cobranding allows for increased revenue, enabling the restaurant operation to attract and hire top-tier management personnel. Furthermore, Hannibal Myers, the esteemed vice president of new business development for Church, highlights the cost-saving benefits of utilising a shared back-of-the-house system.

In conclusion, cobranding is a strategic move that not only amplifies the drawing power of multiple brands, but also paves the way for heightened revenue and operational efficiency. With these advantages in mind, it becomes clear why cobranding is an invaluable tool for businesses looking to thrive in a competitive market.

Business partners have the incredible ability to enhance each other's success by offering different yet complementary products. For instance, the franchisor of Comfort Inn has joined forces with Huddle House to expand their reach by franchising a restaurant adjacent to their hotel. Moreover, Marriott has secured a long-term licensing agreement with Starbucks, allowing them to introduce Starbucks shops, kiosks, and outlets within their hotels. This partnership has proven so successful that in 2001, Hyatt followed suit and signed a similar agreement with Starbucks. Another example of leveraging brand names is demonstrated by Damon's restaurants in Columbus, Ohio, which utilised the renowned Oreo cookies brand to promote their irresistible Damon's Oreo Pie. Michael Branigan, the esteemed vice president of marketing, proudly asserts that the association with the Oreo name instills a sense of trust and credibility in customers. In fact, a compelling study revealed that an astounding 51 percent of restaurant patrons agreed that brand names guarantee consistent quality, with 50 percent of those surveyed willing to invest a little extra for brand-name products.

Hussain et al. (2020) assert that brand equity embodies the remarkable marketing effects or outcomes that befall a product graced with its distinguished brand name, surpassing those that would transpire if the identical product lacked the sheer influence of its brand name.

Summary

In this section, we have thoroughly elucidated the notion of a product, encompassing its core, facilitating, supporting, and augmented components. Furthermore, we have expounded upon the paramount importance of factors such as atmosphere, customer interaction with the service delivery system, customer interaction with fellow patrons, and customer coproduction, all of which necessitate meticulous consideration when crafting a product. Moreover, we have delved into the profound concept of branding, unveiling the essential conditions that lend unwavering support to the establishment of a brand.

Activity

1. Define products from the H&T sector's perspective and provide some examples.
2. Explain the four levels of products that managers in the hospitality industry must think of when designing their product offers.
3. Explain the concept of branding and its importance in the hospitality industry.
4. Briefly explain the five conditions that a manager in the hospitality industry must consider when taking branding decisions.

References

Hussain, I., Mu, S., Mohiuddin, M., Danish, R. Q., & Sair, S. A. (2020). Effects of sustainable brand equity and marketing innovation on market performance in hospitality industry: Mediating effects of sustainable competitive advantage. *Sustainability, 12*(7), 2939.

Kotler, P. (2012). *Kotler on marketing*. Simon and Schuster.

Venkatesh, A., Joy, A., Sherry, J. F., Jr., & Deschenes, J. (2010). The aesthetics of luxury fashion, body and identify formation. *Journal of Consumer Psychology, 20*(4), 459–470.

Ward, M. K., & Dahl, D. W. (2014). Should the devil sell Prada? Retail rejection increases aspiring consumers' desire for the brand. *Journal of Consumer Research, 41*(3), 590–609.

Chapter 13

New Product Developement

Chapter Outline

- Introduction
- The Process of Developing New Products
- Generation of Idea
- Screening the Idea
- Developing the Idea into Concept
- Testing the Concept
- Developing a Marketing Strategy
- Developing the Product
- Testing the Market
- Commercialisation of the Product

Objectives

After reading this chapter, you should be able to:

- Identify the process of new product development.
- Explain how ideas for new products are generated.
- Outline the stages of new product development.

Introduction

A company must excel in the art of creating innovative products. Additionally, it must possess adeptness in effectively managing these

DOI: 10.4324/9781032688497-13

products amidst the ever-evolving preferences, technologies, and competitive landscape. Similar to the life cycle of humans, every product undergoes distinct phases from its inception, progresses through various stages, and ultimately ceases to exist as newer products emerge to better cater to consumer demands. The forthcoming chapter will delve into the exploration of how new products are conceived and the intricate processes or stages that a product traverses during its development.

The product life cycle poses two prominent challenges. Firstly, given that all products ultimately dwindle, it becomes imperative for a firm to discover novel products to supplant the aging ones (the challenge of new product development). Secondly, the firm must possess a comprehensive understanding of how its products mature and adapt marketing strategies accordingly as these products journey through the different stages of their life cycle. We shall first examine the issue of identifying and cultivating novel products, followed by an examination of the challenge of effectively managing them throughout their life cycles.

It is crucial for all organisations in the hospitality industry to establish a dedicated programme for the development of new products. It is necessary for every organisation to establish a programme specifically designed for the development of fresh products. An expert in the field estimates that around half of the profits generated by all American companies come from products that did not exist ten years ago.

A company can obtain innovative products through two separate channels. The first entails the procurement of an entire company, a patent, or a license allowing to produce a product created by someone else. Given the rising costs associated with the development and introduction of significant new products, many companies choose to acquire established brands rather than creating new ones. This is exemplified by Accor's acquisition of Motel 6, McDonald's acquisition of Boston Market, Onomo Hotels' acquisition of Bon Hotels, Choice's acquisition of Rodeway, Econo Lodge, and Friendship Inns, as well as Carnival's acquisition of Holland American, Windstar, Costa, and Seabourn. A company has the option to acquire new products through the establishment of its own research and development department. These recent items pertain to the establishment of innovative items, enhancements to current items, alterations to current items, and the formation of new trademarks that the organisation engages in via its own research and development (R&D) efforts. The focus of this chapter centres around the process of new product development.

The Process of Developing a New Product

1. Generating an Idea

New product development commences with the initiation of idea generation, which entails a meticulous exploration for novel concepts. It is not important that a company generates many ideas in order to identify the valuable ones among them. To make sure the success of new product development, it is recommended to embrace a methodical approach for the pursuit of fresh product ideas rather than relying solely on chance occurrences. Alternatively, the company may come across ideas that do not align with its unique business model. The presence of innovative and creative ideas serves as an indispensable catalyst in the process of generating original concepts. In order to enhance the generation of creative ideas, it is advantageous to provide stimulating influences to a proficient design team (Sozo & Ogkiari, 2019).

A company must meticulously outline its novel product development strategy. This strategy must clearly specify the products and markets that deserve the utmost emphasis. Moreover, it must articulate the company's desired outcomes from these new products, be it an influx of cash, an expanded market share, or any other objective. Take, for instance, the case of McDonald's, which introduced salads to counter the threat of losing market share to Wendy's salad bar. Similarly, Pizza Hut introduced individual pizzas to entice lunchtime customers. Lastly, the strategy must explicitly state the level of dedication to be invested in developing original products, revamping existing ones, and imitating competitors' offerings.

To acquire a constant stream of fresh product ideas, the company must tap into various sources of inspiration. The subsequent section delves into the primary sources of these innovative product ideas.

Internal Sources

A company must meticulously outline its novel product development strategy. This strategy must clearly specify the products and markets that deserve utmost emphasis. Moreover, it must articulate the company's desired outcomes from these new products, be it an influx of cash, an expanded market share, or any other objective. Take, for instance, the case of McDonald's, which introduced salads to counter the threat of losing market share to Wendy's salad bar. Similarly, Pizza Hut introduced individual pizzas to entice lunchtime customers. Lastly, the strategy must explicitly state the level of dedication to be invested in developing original products, revamping existing ones, and imitating competitors' offerings.

To acquire a constant stream of fresh product ideas, the company must tap into various sources of inspiration. The subsequent section delves into the primary sources of these innovative product ideas.

In the hotel industry, the responsibility of making new product decisions is carried out by individuals at both the corporate and property levels. It should be noted that these decision-makers at the corporate level are individuals of high standing, ranging from mid-level to top management personnel. It is noteworthy to mention that in particular circumstances, individuals who are not directly affiliated with the company but have a close association, such as financiers, attorneys, and advisors, actively participate in this intricate process.

When it comes to the property level, the responsibility of making new product decisions usually falls upon the owner, particularly if the hotel is not owned by a prominent chain. In certain cases, the proprietor may choose to delegate this responsibility to a representative, such as a leader, who acts on their behalf. Other prominent figures who play a crucial role in this process include the revered general manager, department managers, and directors overseeing various departments. Moreover, it is not uncommon for a corporate vice president from the chain to actively partake in this decision-making process, further enriching the collective expertise and wisdom.

Customers

Observing and actively listening to customers contribute to the genesis of almost 28 percent of new product concepts. Effectively uncovering consumer requirements and wants can be achieved through the utilisation of consumer surveys. By carefully scrutinising customer inquiries and grievances, the company can uncover innovative products that more effectively address consumer issues. Company management or sales representatives can engage in meaningful discussions with customers to gather invaluable suggestions. The frontline employees, who regularly interact with customers, possess remarkable ability to discern customer needs through their interactions. Managers acquire profound insights into the requirements of guests by personally traversing the premises of the hotel or restaurant and engaging in conversations with customers. Moreover, consumers themselves often create ground-breaking products. Companies can greatly benefit by identifying and introducing these products into the market. As an example, Pillsbury consistently acquires outstanding new recipes through its eagerly awaited yearly Bake-Off event. Interestingly, one of Pillsbury's four cake-mix product lines and several variations of another line can trace their origins directly back to the winning recipes from the Bake-Off competition. The Country Inn proprietor graciously organised a splendid dinner gathering exclusively for her esteemed companions. With great enthusiasm, she requested that each guest bring their most cherished culinary creations, accompanied by an ample supply of recipe copies to be shared amongst all attendees. This delightful occasion not only provided an evening filled with captivating conversations but also served as a wellspring of inspiration for future menu enhancements at the esteemed Country Inn.

Numerous distinguished establishments of opulence habitually arrange a weekly soiree dedicated to pampering their esteemed clientele. The astute general manager, along with the proficient heads of various departments, gracefully assume the role of gracious hosts during these refined affairs. This strategic arrangement directly facilitates management's ability to engage the guests in informal discussions, thereby gaining invaluable insights and recommendations on how to consistently deliver exceptional service. Considering the fact that these esteemed guests are well-versed in the world's most luxurious hotels and are respected leaders in their respective fields, their valued opinions and innovative ideas are highly regarded and treasured.

Competitors

Approximately 27 percent of new product concepts are derived from the meticulous analysis of rival products. Numerous companies acquire competing new products, meticulously examine their manufacturing process, scrutinise their sales performance, and meticulously evaluate whether it would be advantageous to introduce their own new products. Furthermore, companies can diligently study their competitors' advertisements and other forms of communication to glean valuable insights into potential new products. However, it is imperative that when adopting a competitor's idea, one must execute it with utmost precision and finesse, striving to surpass the original creator. Customers will inevitably compare the imitated version with the original, and any unfavourable comparison could potentially have detrimental consequences for the product.

One can effectively cultivate an aura of innovation by drawing inspiration from other markets. A notable example is Michael Turback, formerly of

Turback's Inn in Ithaca, New York, who actively sought out opportunities to attend the prestigious restaurant show in Chicago and explore various establishments in New York City to gather fresh ideas for his own restaurant. This proactive approach not only positioned him as a visionary in the eyes of his customers but also fostered a culture of continuous improvement and creativity.

Many international entrepreneurs have attempted to replicate the success of various hospitality products. However, these imitation products often pale in comparison to the original, resulting in a tarnished reputation for the entire product category. Consequently, when the authentic company eventually enters the market, it faces the daunting task of overcoming this negative perception. On the other hand, there are instances where foreign companies have managed to create such exceptional products that they establish the benchmark for their respective industries. An exemplary illustration of this can be seen in the renowned restaurant chain Pollo Campero from Central America, which has established the benchmark against which competitors like KFC are evaluated.

When companies decide to adopt ideas from different regions, it is essential for them to carefully consider the cultural and social disparities specific to each region. For instance, a previous chef from a restaurant located on the California coast, renowned for its exceptional cuisine served in a relaxed setting, decided to replicate the restaurant's concept in Bryan, Texas. The California establishment provided high-quality food at a reasonable price. Regrettably, the preferences and spending habits of the residents in the small Texas town differed significantly. Instead of elegantly prepared seafood, the people of Bryan favoured fried seafood or a hearty steak and potato. What was considered reasonably priced in California turned out to be quite expensive in Bryan. Consequently, the chef's sophisticated seafood restaurant had to cease its operations, and the premises are now occupied by a steak house.

Distributors and Suppliers

Distributors, much like tour operators, possess close proximity to the market and have the capability to convey valuable information regarding consumer issues and potential product opportunities. In the case of a travel agent who frequently receives requests for a honeymoon travel package to a specific destination, the said agent can relay this pertinent information to the tour operator, who will subsequently pass it along to the destination in question.

Suppliers, on the other hand, have the ability to inform the company about innovative ideas, methodologies, and materials that can be utilised in the creation of new products. Additionally, suppliers can provide insights on the popularity of food products within competitive restaurants and the selection of new products requested by hotels. For example, a supplier of wine to a bar can offer guidance on the types of wine that are currently in high demand.

Hospitality booths at industry trade shows, seminars, and conferences are frequently sponsored by distributors and suppliers. It is highly recommended to take the opportunity to visit these suites in order to gather valuable insights on current trends and competitive strategies, as well as to establish connections with influential individuals.

Other Sources

Other potential sources of ideas include trade magazines, exhibitions, and seminars; governmental organisations; consultants specialising in new product development; advertising agencies; market research companies; academic and private research facilities; as well as inventors.

2. Idea Screening

The objective of the process of generating ideas is to foster the creation of a significant number of innovative concepts. The purpose of the subsequent stages is to gradually decrease the quantity of ideas, which is done with great care and consideration. The initial phase, known as idea screening, is of utmost importance. The objective of this stage is to swiftly identify and discard ideas that may not meet the desired standards, while also identifying ideas that show great promise. As product development costs escalate during later stages, it becomes imperative for the organisation to proceed only with ideas that have the potential to generate profitable products. A practice that is uncommon among most companies involves mandating their executives to document new product concepts in a standardised format. This procedure enables a comprehensive and equitable evaluation by a dedicated committee responsible for scrutinising new products.

Within this documentation, executives furnish a comprehensive description of the product, meticulously define the target market, and conduct a thorough competitive analysis. Additionally, they provide estimated figures for market size, product pricing, development timelines and expenses, manufacturing costs, and the anticipated return on investment. Moreover, they consider the following inquiries: Does this concept offer advantages specific to our organization? Does it align with the company's goals and strategies? Do we possess the requisite personnel, expertise, equipment, and resources to ensure its success?

Several companies have successfully implemented structured mechanisms for the examination and assessment of new product proposals.

Moreover, they provide approximate figures regarding the size of the market, pricing of the product, timelines and costs for development, expenses for manufacturing, and the anticipated return on investment. Furthermore, they investigate the following queries: Does this concept provide advantages to our distinct organisation? Does it correspond to the aims and strategic course of the company? Do we possess the necessary personnel, expertise, equipment, and resources to ensure its success? Several companies have successfully implemented systematic procedures for the assessment and evaluation of innovative product proposals.

The surviving ideas need to be further expanded upon in order to develop them into product concepts. Ensuring a clear distinction between a basic product idea, a product concept, and a product image is unquestionably crucial. A product idea essentially represents a mere possibility of a product that could potentially be brought into the market by those highly esteemed company managers. On the contrary, a product concept takes it a step further by providing an even more intricate portrayal of the idea, using language that is easily comprehensible to those esteemed consumers. In essence, a product's image encompasses the manner in which consumers visualise an actual or potential product, which is of utmost significance.

Major restaurant chains may find it challenging to implement an untested menu across all of their establishments. Burger King, along with other restaurant chains, effectively utilises the strategy of establishing test market locations in specific cities. In the case of American fries, the Piedmont region of North Carolina was deliberately selected as a test market. Regrettably, the product did not fare well and consequently, it was removed from the menus. It is a common practice for hotels to introduce new product ideas on selected floors and properties. This approach enables them to concentrate their resources on the company's core business, which is lodging, while simultaneously developing a new product.

3. Concept Development

During the late 1970s, Marriott had a wonderful opportunity to understand that their urban hotels were becoming extremely popular. The demand for their hotels was so high that it became challenging for them to meet the needs of all their guests. Therefore, they decided to come up with a brilliant plan to generate more income in secondary and suburban areas. They decided to focus on their primary business – providing comfortable accommodation. This led them to embark on the creation of a fantastic new product. But let's be clear, this new product idea doesn't mean that customers are actually buying anything. No, no, no! What customers really want is something they can touch and feel. So, it's the marketer's obligation to take this idea and come up with different product concepts (because evidently, one is not sufficient), assess if customers will appreciate them, and ultimately select the optimal choice (because we all are aware they are incapable of making decisions independently).

The new product named Courtyard by Marriott was conceptualised. To ensure efficient development, Marriott carefully selected individuals from various departments within the company to oversee the process. Extensive analysis of competitors and the market was conducted, leading to the establishment of the following framework for the project:

1. The product would have a strong focus on catering to the transient market.
2. The number of rooms would be kept to a maximum of 150.
3. Creating a residential atmosphere was a key objective, as Marriott discovered that there was a significant group of hotel users who expressed a preference for homelike environments.
4. Efforts were made to minimise any negative impact on Marriott's existing hotels.
5. The inclusion of a limited-menu restaurant was considered.
6. The availability of public and meeting spaces would be subject to certain limitations.
7. The product would adhere to standardised specifications and be limited to a range of five to eight establishments within a given region.
8. To enhance brand recognition and benefit from a positive association, the Marriott name would be prominently associated with the product. Courtyard by Marriott underwent rigorous concept testing prior to its introduction to the market.

4. Concept Testing

Concept testing is carried out among a carefully chosen group of target consumers, as if they are some special group. New product ideas can be presented in a way that even the most clueless individuals can understand, either by using words or pictures. Marriott, being the genius that they are, used a fancy statistical method called conjoint analysis to test their concept for the Courtyard Motel. They went ahead and showcased different motel configurations to potential target guests, as if these guests couldn't figure it out themselves, and had the audacity to ask them to rank these configurations based on how much they wanted them. As if their opinion really mattered! Then they took these rankings and subjected them to some fancy statistical analysis, because apparently, they couldn't just use common sense, to determine the most optimal motel configuration. Because why trust the judgment of actual guests when you can rely on numbers?

However, in many cases, simpler consumer attitude surveys are utilised. For example, let's assume that 10 percent of the consumers expressed a definite intention to purchase, while an additional 5 percent indicated a probable intention. The company would then extrapolate these figures to estimate the sales volume within the population size of this target group. Nevertheless, it is important to acknowledge that this estimate may be subject to uncertainty, as individuals do not always follow through with their stated intentions.

5. Marketing Strategy

The subsequent steps involve crafting a marketing strategy, as if this should be common knowledge. Within this strategy, we formulate an initial plan, because apparently, that's a necessity, to introduce the product to the market. This marketing strategy statement, which I'm sure you're eager to hear about, consists of three important components.

The first component, which I'm sure holds great significance, entails delineating the target market, defining the intended positioning of the product, and of course, outlining the sales, market share, and profit goals for the initial years. For Courtyard by Marriott, if that's of any relevance, the target markets encompass business travellers seeking reasonably priced yet high-quality accommodations, because apparently, that's a prevalent demand, and leisure travellers in search of secure and comfortable lodging.

Proceeding to the second component of the marketing strategy statement, it precisely outlines the carefully planned pricing, distribution, and marketing budget for the product's first year. Marriott accomplished this by utilising advanced statistical software, which facilitated the creation of intricate models. These models provided invaluable insights into pricing strategies and presented market share projections based on those prices. The segmentation data derived from this process equipped Marriott with the essential information necessary for the effective promotion of their hotels.

Finally, the third segment of the marketing strategy statement explores the enduring sales and profit goals, along with the long-term marketing-mix strategy.

6. Business Analysis

After management reaches a decision about the product concept and marketing strategy, it is absolutely possible to confidently evaluate the business viability of the proposal. To conclusively determine their alignment with the company's objectives, conducting a meticulous examination of sales, costs, and profit projections is necessary. Once these factors have been proven entirely satisfactory, the product can confidently proceed to the product development stage.

To accurately measure potential sales, it is highly recommended for the company to thoroughly explore the sales history of similar products and also conduct an extensive market survey to gather valuable insights. Additionally, estimating both the minimum and maximum sales figures would be exceptionally wise in order to gain a comprehensive understanding of the range of associated risks.

Following the completion of the sales forecast, management can confidently proceed to estimate the anticipated costs and profits for the product. These estimations are carefully calculated with the active involvement of various departments, including Research and Development, Operations, Accounting, and Finance. The analysis also includes a thorough evaluation of marketing expenditures. Subsequently, the company confidently utilizes the sales and cost data to assess the financial viability of the new product.

7. Product Development

If the product concept successfully passes the business assessment, it will move forward to the product development stage. In this stage, the product concept will be transformed into an actual product prototype. Up until now, it has only existed as a verbal description, a visual representation, or a model. This particular stage requires a significant increase in investment and is a crucial point for determining if the product idea can become a viable and practical product. The company will create one or more physical versions of the product concept, with the goal of finding a prototype that meets the following criteria:

- Consumers must perceive it as possessing the fundamental attributes that provide value to them.
- The performance is absolutely secure during typical usage.
- Can be produced within a budgeted cost.

One challenge associated with the development of a prototype is that it often encompasses only the fundamental elements of the product. Numerous intangible factors, such as the performance of the employees, are unable to be incorporated. In the case of a hotel, the prototype is typically constrained to the essential room amenities and services that align with the overall concept. It is imperative for marketers to bear in mind that they must endeavour to provide prospective customers with an understanding of the intangible aspects of the product, which include the provision of supporting and facilitating goods and services.

8. The Market Testing Stage

When the product passes both functional and consumer tests, the next crucial step to consider is market testing. Market testing is a pivotal phase where both the product itself and its accompanying marketing programme are introduced into practical and real-world market environments.

Market testing offers an incredible opportunity for marketers to acquire invaluable experience in effectively promoting the product, identify any potential issues that may arise, and determine areas where additional information is needed before committing to the significant expense of a full product launch. Throughout market testing, both the product and its complete marketing programme undergo a comprehensive evaluation in real market scenarios. This evaluation covers various aspects such as the product's positioning strategy, advertising efforts, distribution channels, pricing structure, branding approach, packaging design, and budget allocation. Moreover, market testing enables the company to assess the response of consumers and dealers in terms of product handling, usage, and repurchasing tendencies. The insights obtained from market testing can subsequently be utilised to enhance sales and profit forecasts, resulting in more accurate projections.

9. Commercialisation

Market testing provides the management with the necessary information to reach a final determination regarding the potential launch of the new product. In the event that the company proceeds with commercialisation, it will encounter considerable expenses. Specifically, it may be required to allocate between $10 million and $100 million solely for advertising and sales promotion within the initial year. As a case in point, McDonald's Corporation, renowned for its signature Big Mac and iconic Golden Arches, initiated the testing of two four-star hotels in Switzerland. Should this endeavour prove successful, McDonald's would subsequently expand its product offering to include Golden Arch Hotels throughout Europe. McDonald's firmly believes that European patrons will value the same attributes in hotels that they have come to expect from their hamburgers: quality, service, cleanliness, and value. In order to successfully launch a new product, the company must make a series of critical determinations: when to launch, where to launch, to whom it should be marketed, and how it should be introduced.

The initial determination revolves around the appropriateness of the moment to introduce the novel merchandise. In the case of Marriott, the hotel that was utilised as a testing ground had the privilege of witnessing a commendable 90 percent occupancy rate.

The organisation faces a crucial choice regarding the implementation of the groundbreaking product, deliberating on whether to release it in one location, a specific region, multiple regions, the local market, or the global market. It is a rare occurrence for companies to have the confidence, financial resources, and capability needed for the nationwide distribution of new merchandise. Instead, they choose to gradually implement a well-planned strategy to introduce the product to the market. Specifically, smaller companies tend to carefully choose an attractive city and carry out an extensive promotional campaign to enter the market. Eventually, they may proceed to enter other cities, one by one. Larger companies may decide to begin the introduction of their product in a specific region, followed by subsequent expansion into the next region. Companies with well-developed national distribution networks, such as those in the automobile industry, often initiate the introduction of their new models in the domestic market. After careful deliberation, Marriott has decided to introduce the Courtyard concept into regional markets comprising a selection of five to eight locations. As of January 1986, the company had effectively established 300 sites that were either operational, under contract, or in the midst of construction.

In order to effectively penetrate the rollout markets, it is imperative for the company to strategically focus its distribution and promotional efforts towards the most suitable target groups. Prior to this stage, management should have conducted thorough market testing to identify the profiles of the most promising prospects. At this moment, the company must improve its market identification process, diligently seeking out early adopters, heavy users, and influential opinion leaders.

Moreover, the company has to develop a precise action plan for the introduction of the new product into the selected markets and allocate the marketing budget strategically across the marketing mix.

Summary

In this chapter, we discussed how new product ideas are generated and the stages of new product development.

Activity

1. Explain the sources of ideas for generating new products in the hospitality industry and how important each source is.
2. Describe the new product development stages for a hospitality firm?
3. Explain the importance of the stages involved in developing a new product or service for the managers in the tourism industry in Africa?

Reference

Sozo, V., & Ogliari, A. (2019). Stimulating design team creativity based on emotional values: A study on idea generation in the early stages of new product development processes. *International Journal of Industrial Ergonomics*, *70*, 38–50.

Chapter 14

The Product Life Cycle

Chapter Outline

- Introduction
- Product Life Cycle
- Introduction Stage
- Growth Stage
- Maturity Stage
- Decline Stage

Objectives

After reading this chapter, you should be able to:

- Describe the various stages a product goes through in its lifetime.
- Examine the implications of each stage of the product life cycle for marketing.
- Understand strategies that managers could implement at the stage of the product's life to maximise its life.

Introduction

After the successful launch of a new product, it is vital for management to ensure that the product enjoys a prolonged and highly profitable existence.

DOI: 10.4324/9781032688497-14

While it is recognised that the product may not have everlasting sales, it is essential for managers to generate enough profit to offset the costs incurred during the development and marketing phases, as well as for the efforts and risks involved. In order to truly optimise profits, it is customary for a product's marketing strategy to be reevaluated and reworked multiple times. These strategic adjustments are frequently driven by changes in the market and environmental conditions that arise as the product progresses through its life cycle. Therefore, it is crucial to analyse how the hospitality and tourism sectors are impacted by the different stages of the product life cycle.

Product Life Cycle Strategies

The journey of a product's existence is marked by five distinct stages that hold immense significance. These stages, specifically product development, introduction, growth, maturity, and decline, encapsulate the essence of a product's entire life cycle.

The genesis of a product begins with the diligent efforts of a company to discover and nurture a novel product idea. During the phase of product development, sales may not be visible, yet the company's unwavering investment costs continue to accumulate, fueling the fire of innovation.

The opening stage of a product's life cycle is distinguished by a period of gradual sales growth as the product is introduced into the market. It is during this pivotal time that profits may not yet be in sight, for the expenses incurred during product introduction are substantial and demand undivided attention.

Growth is an exhilarating phase where the market embraces your product with open arms, resulting in an exponential surge in profits.

Maturity marks a slight deceleration in sales growth, as your product has already conquered the hearts of its potential buyers. Profits stabilise or slightly decline due to amplified marketing investments that are required to safeguard your product against fierce competition.

Decline signifies a swift decline in sales and a corresponding drop in profits, urging you to take immediate action. Not all products adhere to the traditional S-shaped product life cycle. Some products are introduced quickly and come to an untimely end. The concept of the product life cycle is extremely adaptable and can be used to describe a wide range of things, whether it is a category of products (like fast-food establishments), a style of product (such as fast-food burgers), or even a specific brand (like Wendy's).

However, it is crucial to acknowledge that the product life cycle appears differently depending on the context.

Product categories, for example, typically have the longest life cycles, with many remaining in the mature stage for a significant period of time. Conversely, product types often follow the standard path of the product life cycle. Take, for instance, the drive-in restaurant or the roadside tourist court, both of which experience a predictable sequence of events, including introduction, rapid growth, maturity, and eventual decline. The life cycle of a particular brand can rapidly fluctuate due to the ever-changing landscape of competitive challenges and responses. It is important to note this.

The notion of the product life cycle (PLC) provides a priceless framework for grasping the dynamics of products and markets. However, when it comes to forecasting product performance or devising marketing strategies, the employment of the PLC concept presents several practical challenges. For instance, managers may confront obstacles in identifying the current life-cycle stage of a product, precisely determining the moment it transitions to the next stage, and comprehending the myriad of factors that influence its progression through these stages. In practical terms, predicting sales levels at each PLC stage, estimating the duration of each stage, and determining the shape of the PLC curve can be a complex and demanding task.

Most marketing texts highlight the importance of the PLC, yet astonishingly few managers acknowledge its value in shaping their marketing strategies. This can be attributed to two plausible explanations. To begin, managers typically base their strategic choices on a thorough evaluation of the distinct attributes associated with each stage of the product life cycle, even when they don't explicitly rely on the product life cycle as their guiding framework. Secondly, it's important to recognise that making accurate forecasts regarding the trajectory of the product life cycle can be a formidable challenge. Many products veer away from the standard path, making it exceptionally difficult to generate precise forecasts.

It is imperative to recognise that the product life cycle should not be perceived as a simplistic tool for predicting the lifespan of a product. Instead, it functions as a formidable framework for understanding the influence of market dynamics, environmental factors, and competition, thereby enabling a deeper comprehension of how a product might react to different stimuli. By embracing the notion that products inherently possess life cycles with distinguishable stages, valuable insights can be gleaned on how to effectively manage and prolong the life of a product. Unmanaged products effortlessly navigate through the various stages of their life cycle, encountering minimal resistance. Environmental and competitive changes serve as catalysts that propel a product along its life cycle, demanding companies to swiftly respond in order to ensure the continued marketability of their products.

Using the product life cycle (PLC) concept to formulate an influential marketing strategy can potentially pose certain obstacles. The plan acts as both the cause and effect of a product's life cycle: the current stage of the product in its PLC provides valuable insights for determining the most effective marketing strategies, while subsequent marketing strategies directly influence the product's performance in its later life cycle stages. However, when applied thoughtfully, the PLC concept is unlikely to serve as a potent tool for devising successful marketing strategies tailored to each phase of the product life cycle.

Now, let us thoroughly explore the strategies that can be employed for each of the remaining life-cycle stages.

1. Introduction Stage

The initiation phase begins when the new product becomes available for purchase. The initiation requires a substantial amount of time, and the augmentation of sales is inclined to be sluggish. A few commodities may persist in the initiation phase for numerous years before they transition into a phase of expeditious growth; resorts, for instance, adhere to this sequence as it necessitates time for the resort to attain widespread recognition. Numerous enterprises adopt what Theodore Levitt articulates as the "previously tested fruit strategy." They observe others entering the market as trailblazers. Upon the introduction of resorts, numerous tourists refrained from participation until the product confirmed its excellence in the competitive arena. Being a trailblazer entails a degree of uncertainty, yet those who remain on the sidelines may bear witness to others swiftly accumulating market share if the product is highly sought-after. The pioneers are positioned incredibly well to safeguard their dominant market share against any potential threats posed by latecomers. During the initial phase, the profits may be modest or even negative due to the relatively low sales volume and the significant expenses incurred in terms of distribution and promotional activities. It is imperative for a company to have the necessary capital to entice distributors and ensure a smooth flow of products. Substantial resources are allocated towards promotional endeavours, aiming to effectively inform consumers about the innovative nature of the product and actively encourage them to give it a try.

In the initial stage, competition is limited to only a few chosen companies that offer basic versions of the product because the market is not fully ready for more advanced versions. These companies specifically aim at customers who are ready to buy, usually belonging to higher-income demographics. Prices tend to be higher during this phase because of limited production capacity, occasional manufacturing difficulties, increased promotional expenses, and other associated costs.

2. Growth Stage

If the new product successfully meets the demands of the market, it will undoubtedly enter the stage of rapid growth, causing sales to soar. The initial group of trendsetters will continue to make purchases, serving as an influential force for the rest of the buyers to follow suit, especially when they hear positive recommendations from others. In the case of the resort, it will undoubtedly witness a surge in popularity and allure. Competitors, enticed by the lucrative opportunities for profit, will eagerly enter the market. These competitors will introduce innovative product features, ultimately expanding the market even further. The influx of competitors will subsequently lead to a greater number of outlets and a significant leap in sales.

Prices stay at their current levels or experience only a minor decrease. Companies maintain their promotion budgets at the same level or even slightly increase them in order to stay competitive and further educate the market. As a result, profits soar during this stage of growth, as the costs of promotion are distributed across a larger volume, more efficient systems are established, and corporate management expenses are divided among a greater number of units.

The firm employs a variety of strategies to ensure the continuous expansion of the market:

The firm enhances the quality of its products and introduces new features and models to entice consumers.

It ventures into new market segments, exploring untapped opportunities for growth.

It explores new distribution channels to reach a wider audience and expand its market reach.

The firm strategically allocates its advertising efforts, moving away from simply creating product awareness and instead focusing on building strong conviction and encouraging purchase.

At the inappropriate moment, the firm strategically elevates prices to discourage a larger pool of buyers.

In the growth stage, the company must make a crucial decision to either strive for a substantial market share or maximise their current profits. By making substantial investments in product improvement, marketing, and distribution, the company establishes itself as a prominent competitor. Even though this approach may require sacrificing immediate profits, the company expects to regain these losses in the following stages.

3. *Maturity Stage*

At a certain point in time, a product undergoes a slowdown in sales growth, indicating its entry into the maturity stage. In the case of the resort, its initial appeal might have decreased, and tourists might have begun exploring other destinations. This phase typically lasts longer than the previous two stages and poses significant challenges for marketing management. The majority of products find themselves in the maturity stage of their life cycle, and as a result, marketing management primarily focuses on mature products.

The decline in the rate of sales growth leads to a scenario where the supply surpasses the demand. As a result, an excess capacity emerges, leading to intensified competition. Rival companies commence a process of reducing prices, while simultaneously enhancing their advertising and sales promotions. The emergence of "Burger wars" and "pizza wars" can be attributed to the fact that these products have reached a mature stage. In this stage, the actual growth in sales for a particular product category is roughly equivalent to the growth of the population. The sole means of significantly boosting sales is by enticing customers away from the competition. Consequently, battles over pricing and intensified advertising campaigns serve as the primary methods employed for this purpose, both of which ultimately lead to a decline in profit. Less powerful competitors are gradually eliminated from the market, leaving only well-established rivals in the primary market segments. At the same time, smaller competitors focus on targeting specific niche markets.

Having a robust defense is the ultimate way to protect oneself. The product manager must not only protect the product but also actively search for opportunities to enhance the target markets, refine the product, and optimise the marketing mix.

Market Modification

At this crucial stage, the proactive product manager commits significant efforts to enhance product consumption. With unwavering determination, the manager actively pursues new users and explores untapped market segments, all with the goal of increasing usage among existing customers. McDonald's, in a strategic move to attract new customers and improve utilisation, introduced a range of enticing breakfast options, appealing salads, irresistible desserts, and delicious chicken sandwiches. Moreover, product managers possess the skill to effectively reposition the brand, catering to a wider audience or a rapidly expanding segment. For example, when anti-drunk driving campaigns successfully reduced the consumption of alcoholic beverages, Bennigan's smartly shifted its focus to highlight its delectable cuisine, successfully transforming its image from a mere drinking establishment into an exquisite dining destination.

Product Modification

'The product manager can modify various aspects of the product, such as its characteristics, quality, features, or style, with the intention of captivating new users and igniting a stronger desire for usage among consumers. The strategic modification of the product provides a valuable opportunity to align it more closely with the customer's deepest desires, core values, and profound interests. Taking advantage of the surge in popularity of low carbohydrate diets due to heightened health consciousness, McDonald's has skillfully introduced Keto diets to its menu.

Product modification is a strategic approach that focuses on enhancing the various aspect of the product, like durability, reliability, speed, and taste to improve its overall performance. This approach proves fruitful when improvements in quality are noticeable to buyers, instilling a belief in the claim of enhanced quality.

Marketing-Mix Modification

The product manager can enhance sales by making changes to one or more elements of the marketing mix. In order to attract fresh users and customers from competitors, prices can be reduced. An improved advertising campaign can be devised to create a stronger impact. The utilisation of aggressive sales techniques, promotional trade deals, price reductions, giveaways, and engaging contests can also be employed. Additionally, the company can expand its presence in larger market channels, particularly if these channels are experiencing growth. Furthermore, the company can offer novel or enhanced services to prospective buyers. For instance, recognising the significance of social media, Meliá Hotels International, the third-largest hotel group in Europe, revolutionised its promotional strategy by partnering with Hootsuite to make social media a central component of its new promotional approach.

4. Decline Stage

The sales of most products and brands inevitably decreases over time. This decrease can occur gradually, as demonstrated by the decline of Steak and Ale, or rapidly, as exemplified by the downfall of Victoria Station. The sales can either experience a sharp decline to zero or reach a low level and persist there for an extended period. For instance, the Victoria resort can no longer compete with more modern and innovative establishments.

A number of elements are involved in the decline of sales, including advancements in technology, shifts in consumer preferences, and increased competition. Consequently, as sales and profits diminish, certain companies choose to withdraw from the market altogether. Those that decide to stay may opt to reduce the range of products they offer. They might eliminate smaller market segments and less profitable trade channels. Additionally, they may decrease their promotional budget and further reduce their prices.

Maintaining a weak product can be extremely detrimental to the organisation, not only in terms of diminished profits, but also due to the presence of concealed expenses. A product at the decline stage does not only require excessive attention from management, but also necessitates frequent adjustments in terms of pricing and inventory. The resources and efforts invested in advertising and sales force for such a product could be better utilised to enhance the profitability of superior products. Furthermore, the tarnished reputation of the weak product can erode customer confidence not only in the company itself, but also in its other offerings. However, the most significant cost may lie in the future. By persisting with weak products, the organisation delays the search for suitable replacements, leading to an imbalanced product mix that adversely affects current profits and weakens the company's position in the future market.

Companies must prioritise their aging products for several compelling reasons. It is crucial for them to consistently monitor and evaluate the sales, market shares, costs, and profit trends of each product. This strategic approach will effectively pinpoint products that are currently in the declining stage.

For every product that is experiencing a decline, the management is faced with the crucial task of making a decision: Should they maintain it, harvest its potential, or drop it altogether? In the case of harvesting, the management has the opportunity to reduce various costs associated with the product. A prime example of this can be seen with Steak and Ale, as they closed down some of their less profitable restaurants that offered reel meat products, while still maintaining their more lucrative locations. They even went as far as developing new restaurants in areas where customers still craved a mouthwatering steak when dining out. Ultimately, this calculated step aims to decrease the company's losses in the short term. On the flip side, leadership might select to remove the product from their line entirely. This can involve selling it to another firm or liquidating it at its salvage value. If the company has plans to identify a potential buyer, it would not be in their best interest to exhaust the product's value through gathering.

Product Deletion

The product life cycle presents a captivating case for the inevitability of product obsolescence, the loss of market appeal, and the subsequent need for replacement. Nonetheless, it is essential to acknowledge the potential danger of premature replacement, as introducing a product to the market requires substantial investments of time, effort, and financial resources. It is a widely known fact that merely half of new products actually attain profitability. Accordingly, when a company successfully launches a winning product, it is only normal that they would want to maximise the benefits derived from it. In light of this, prudent management would be hesitant to discontinue a product that still holds the potential for profit. Conversely, if a product reaches a point where it is no longer marketable, it becomes imperative to terminate it rather than persistently allocating valuable time and resources towards reviving it.

According to Zhu et al. (2022), research on product deletion remains relatively scarce and poorly explored. The decision to remove a product from the market often lacks appeal for managers, primarily due to the significant operational changes and disruptions it can cause within the organisation and throughout the supply chain.

Activity

1. What is the product life cycle? Describe each stage of the product life cycle and their implications on the performance of the product in the hospitality industry.
2. How can managers in the hospitality and tourism industry in Africa use the product life cycle to increase brand performance?
3. Explain the strategies that managers should adopt at the product maturity stage PLC to compete effectively.

Reference

Zhu, Q., Kouhizadeh, M., & Sarkis, J. (2022). Formalising product deletion across the supply chain: Blockchain technology as a relational governance mechanism. *International Journal of Production Research*, *60*(1), 92–110.

Chapter 15

Pricing Products: Pricing Considerations, Approaches, and Strategy

Chapter Outline

- Introduction
- Price
- Factors to Consider When Setting Prices
- External Factors Affecting Pricing Decisions
- General Pricing Approaches
- Pricing Strategies

Objectives

After reading this chapter, you should be able to:

- Explain the internal factors that have an impact on pricing decisions, particularly focusing on marketing objectives, the marketing mix strategy, cost considerations, and organisational factors.
- Recognise and clarify the external influences that impact pricing choices, including market dynamics, demand fluctuations, competitive forces, and various elements within the external environment.

DOI: 10.4324/9781032688497-15

- Highlight the distinctions between various pricing methodologies and be capable of discerning the differences among cost-plus pricing, target-profit pricing, value-based pricing, and going-rate pricing.
- Recognise the distinct pricing strategies for new products, namely market-skimming pricing and market-penetration pricing.

Introduction

Marketers and managers must possess a deep understanding of the art of price management. It is of utmost importance to strike the perfect balance, as charging an exorbitant amount will inevitably drive away potential customers, while charging a meager fee may result in a company struggling to generate sufficient revenue to sustain its operations effectively. In the forthcoming section, we shall lightly touch upon the intricate world of pricing, exploring the myriad approaches and strategies that can be adopted by marketers in the hospitality and tourism industry.

Price

Price is the monetary value assigned to a particular good or service. In a broader sense, price represents the cumulative worth that consumers transfer in exchange for the advantages derived from possessing or utilising the said product or service. The amount is a vital component of the marketing mix, playing a crucial role in generating revenue. Pricing and price competition are widely regarded as the most daunting challenge for marketing executives, according to a number of experts. Even though pricing is the least understood aspect of the marketing variables, it can still be controlled in an unregulated market. Frequently, pricing alterations are hastily implemented without undergoing meticulous analysis, serving as mere temporary remedies. The most common mistakes when it comes to pricing are (a) excessive focus on costs, (b) failure to adjust prices according to market fluctuations, (c) lack of consideration for the other elements of the marketing mix, and (d) insufficient variation of prices across different product items and market segments. It is imperative to acknowledge that a single pricing blunder can precipitate the collapse of an otherwise prosperous business. Consequently, every manager must possess a fundamental understanding of pricing principles.

Rent is linked to payment for an apartment, while a hotel sets a rate for an overnight stay. Tuition encompasses the cost of education, whereas physicians or dentists demand fees for their services. Airlines, railways, taxis, and bus companies impose fares for their services. Hotels determine room rates, while banks apply interest charges for the use of their funds. Driving on Florida's Sunshine State Parkway results in incurring a toll fee. Front-desk clerks receive wages, and bartenders earn both wages and tips. Real estate agents who facilitate restaurant sales collect commissions. Lastly, income taxes represent the cost associated with the privilege of earning income.

Marketers and managers must have a deeper and better comprehension of price. Overcharging pushes away potential customers. Undercharging may result in insufficient revenue for an enterprise to adequately sustain its operations. The deterioration of equipment, the staining of carpets, and the need to repaint painted surfaces are all matters that necessitate attention. Ultimately, a company that fails to produce sufficient income to sustain its operations will ultimately cease to exist. Within this chapter, we explore the various factors that must be taken into account by hospitality marketers when establishing prices. Additionally, we delve into general pricing methods, pricing strategies for novel products, product-mix pricing, the initiation and response to price alterations, and the adjustment of prices to align with buyer and situational factors.

Gibbs et al. (2018) posit that Airbnb have a distinctive and intricate framework in relation to dynamic pricing, wherein hosts exhibit a limited utilisation of dynamic pricing tactics, particularly when juxtaposed with hotels.

Mitra (2020) also notes that the extent of asymmetrical impact is considerable with respect to revenue per available room and occupancy rate during a decline in tourist arrivals, in contrast to the corresponding extent observed during an increase in tourist arrivals.

Factors to Consider When Setting Prices

The pricing determinations of a company are swayed by a multitude of factors, both from within and without. From within, these factors encompass the marketing goals, marketing mix tactics, expenses, and organisational deliberations of the company. Conversely, external factors entail the market dynamics, consumer demand, competitive landscape, and diverse environmental aspects.

Internal Factors Influencing Pricing Decisions

1. ***Marketing Objectives***

Before an organisation can establish a price, it must carefully select a product strategy. Once the organisation has made a well-informed choice about its target market and positioned itself strategically, its marketing-mix strategy, including pricing, becomes highly accurate. For example, consider the brand Four Seasons, which positions its hotels as symbols of luxury and charges premium rates for its rooms. In contrast, Motel 6 and Formula One have strategically positioned themselves as affordable motels with limited services, catering to budget-conscious travellers. This specific market positioning naturally requires lower prices. Therefore, the strategic decisions regarding market positioning have a significant impact on determining the price.

According to Lockyer (2005), "The decision concerning room pricing is one of the most crucial aspects of hotel marketing strategies, as the price of a hotel room greatly influences the decisions made when selecting accommodations." The more clearly defined a company's objectives are, the simpler it becomes to establish prices. Common objectives include survival, maximising short-term profits, maximising market share, and achieving product-quality leadership.

Survival: Companies facing challenges such as excessive capacity, intense competition, or shifting consumer preferences prioritise their survival as their primary objective. In the short term, the focus on survival outweighs the pursuit of profit. This strategy is frequently employed by hotels during periods of economic downturn. Similarly, manufacturing firms could potentially opt for lowering production levels to coincide with demand. In the midst of a recession, a hotel with 300 rooms still has all 300 rooms available for sale each night, despite experiencing a decline in demand to only 140 rooms per night. To weather this scarcity, the hotel endeavours to reduce rates and optimise cash flow within the given circumstances. Consequently, this approach inevitably affects not only the immediate competitors but occasionally the entire industry as well. Competitors in the hospitality sector are acutely aware of fluctuations in pricing and typically respond accordingly if they perceive a threat. As a result, there is a noticeable impact on key indicators: occupancy rates decline, room rates decrease, and profits diminish.

Observers of the hospitality industry have occasionally proposed that careful monitoring of competition employing a survival pricing strategy is warranted, albeit not necessarily to be replicated. In the event that the hotel is one of two establishments in a market, such as a small town, the potential impact of reducing prices could be substantial. Conversely, in the case of a hotel situated in Orlando, Florida, where numerous competitors exist, it merely represents a fraction of the overall room supply. Under these circumstances, rivals endowed with robust marketing programmes may be inclined to leverage their marketing prowess to attract clientele rather than resorting to price reductions. Furthermore, for a hotel boasting effective marketing strategies, it may be advantageous to permit a competitor to lower prices and attract budget-conscious customers, thereby reserving the more lucrative business for themselves, especially if the hotel employing the survival strategy possesses a limited market share.

Current Profit Maximisation: Numerous companies strive to establish a price that maximises their current profits. They conduct extensive analyses of projected demand and the costs associated with different pricing options, carefully selecting the one that will generate the highest immediate profit, cash flow, or return on investment. Their primary focus is on attaining immediate financial gains rather than long-term performance. For example, a company may purchase a financially distressed hotel at a reduced price. Their objective is to revitalise the hotel, demonstrate a profitable operation, and eventually sell it. If the hotel owners can successfully accomplish this turnaround, they have the potential to realise a significant capital gain.

Market-Share Leadership: Other companies are eager to acquire a dominant position in the market. They are of the firm belief that by possessing the largest market share, a company can benefit from reduced expenses and substantial long-term profits owing to economies of scale. This, in turn, leads to the establishment of very competitive prices. Marriott, on the other hand, is fully committed to becoming the market leader in its respective field. Whenever a new hotel is launched, Marriott takes swift action to build its market share at an accelerated pace. To illustrate, when Marriott introduced its resort on Australia's Cold Coast, they initially offered rates as low as $99. However, just six months later, they were able to raise the prices to almost double this rate. The low opening rates successfully generated demand, which led to an increase in business. As demand continued to grow, the lower-revenue clientele was gradually replaced with higher-paying customers. This strategy decisively utilises pricing and other components of the marketing mix to establish a convincing perception of providing unparalleled value in comparison to competitors.

Product-Quality Leadership: The Ritz-Carlton chain is well-known for its substantial construction or acquisition cost per room, consistently surpassing $500,000. Furthermore, luxury chains also encounter substantial labour expenses per room. These esteemed establishments provide an all-encompassing service that demands highly qualified personnel and a remarkable employee-to-guest ratio. Consequently, they must set a premium price for their opulent hotel accommodations.

Distinguished industry leaders such as Ritz-Carlton and Four Seasons Hotels and Resorts command higher prices for their offerings, but this is merely a reflection of their unwavering commitment to excellence. They continuously reinvest in their operations, thereby ensuring their unwavering status as paragons of quality.

2. ***Marketing-Mix Strategy***

The marketing mix is unquestionably a powerful set of tools that an organisation skillfully modifies to achieve its intended goals, which can include profitability, market share, customer satisfaction, and overall survival.

Price, one of the tools in the marketing mix, plays a vital role in ensuring a company's marketing objectives are achieved. It is imperative that the price aligns with the product design, distribution, and promotional decisions to create a cohesive and impactful marketing strategy. It is important to note that decisions made regarding other variables within the marketing mix can have a significant impact on pricing decisions. For example, resorts that plan to distribute most of their rooms through wholesalers must incorporate sufficient margins into their room prices to offer substantial discounts to the wholesalers. Given that hotels typically renovate their facilities every five to seven years to maintain a high standard, it is crucial to set prices in a manner that takes into account the costs of future renovations. A firm's promotional mix unquestionably has a substantial impact on pricing, as it undeniably plays an essential role in shaping the value proposition of the product or service. For instance, when a restaurant caters for conventioneers, it may face challenges in building repeat business compared to a neighbourhood restaurant. In such cases, it becomes imperative for the restaurant to invest in targeted advertising through city guides that specifically cater to the needs of conventioneers. Failing to take into account the promotional costs while setting prices can lead to detrimental consequences on the revenue and overall cost structure of the restaurant.

In the realm of business, pricing decisions are often given utmost importance. The price chosen by a company acts as a cornerstone for making other marketing-mix decisions. A striking illustration of this can be seen in the case of Marriott, which identified a lucrative opportunity in the economy market and strategically developed Fairfield Inns. By leveraging the power of pricing, Marriott successfully positioned its motel chain in the market. The target price set for Fairfield Inns played a pivotal role in defining the product's market, assessing competition, shaping its design, and determining the product features. It is crucial for companies to carefully consider all aspects of their marketing mix in a holistic manner when crafting a comprehensive marketing programme that yields optimal results.

3. ***Costs***

Costs establish the foundation for determining the price a company can demand for its product. It is imperative for a company to establish a price that not only encompasses the expenses linked to manufacturing, distributing, and promoting the product but also guarantees a reasonable profit for its investors. As a result, the costs experienced by a corporation play a crucial role in shaping its pricing strategy. Many companies are determined to excel in the production of goods and services at a competitive cost within their respective industries. McDonald's, for example, has developed highly efficient systems for producing fast food. Consequently, a new hamburger franchise would face significant challenges in trying to compete with McDonald's in terms of cost. Successful low-cost producers achieve substantial cost savings through enhanced efficiency while upholding product quality. Companies that are able to reduce their costs can offer lower prices, thereby capturing a larger market share. However, it is important to note that lower costs do not always result in lower prices. Some companies with low costs opt to keep their prices at the same level as their competitors, thus providing a higher return on investment.

External Factors Affecting Pricing Decisions

The inherent characteristics of the market and its demand, the competitive landscape, and various other environmental factors are external factors that undeniably influence pricing decisions.

1. **Market and Demand**

While costs establish the minimum thresholds for pricing, it's the market and demand that set the upper limits. Both consumers and distribution channel buyers, like tour operators, scrutinise a product's price in relation to the benefits it provides. Therefore, before setting prices, a marketer must have a deep understanding of the relationship between price and demand for a particular product.

2. **Cross-Selling and Upselling**

The owner of La Colombe d'Or has strategically employed the highly effective tactic of cross-selling, a fundamental component of successful revenue management. The hospitality industry presents abundant opportunities for cross-selling, showcasing its remarkable potential. For example, a hotel can adeptly cross-sell various services such as access to an exercise room and executive support services like faxing. It can also offer an enticing array of retail products, ranging from meticulously handcrafted chocolates to luxurious terrycloth bathrobes. Likewise, a ski resort can intelligently cross-sell exhilarating ski lessons and enchanting dinner sleigh rides. Hilton Worldwide, a prominent hospitality brand, takes it a step further by providing an exceptional digital check-in service that empowers guests to personally choose desirable upgrades and conveniently request additional amenities before arriving at the hotel.

Upselling is an integral and effective management technique of maximising performance. This valuable technique is achieved by providing comprehensive training to sales and reservations personnel, instilling in them the continuous motivation to offer higher-value priced products rather than settling for lower prices. Among the various established strategies for upselling, one common approach is the concept of room upgrade packages to esteemed guests. During the process of check-in for instance, the astute front desk staff effortlessly presents the option of ascending to a more spacious or opulent accommodation, thereby justifying a higher tariff. It is indisputable that this method significantly augments the hotel's financial gains. Furthermore, hotels adeptly employ the upselling technique by extending the courtesy of early or late check-ins to their esteemed guests. A respected advocate of this method firmly believes that any hotel, regardless of its calibre, has the potential to increase its catering revenue by a staggering 15 percent through the intelligent practice of upselling.

General Pricing Approaches

Determining prices for products and services is not merely an important managerial practice but also a critical organisational capability that profoundly affects a firm's reputation, growth, and profitability (Dutta et al., 2003; Johansson et al., 2012).

Determining the price of a product or service necessitates discovering a delicate balance between a price that is too low to yield a profit and one that is too high to generate demand. While product costs provide a baseline price, it's the consumers' perception of the product's value that sets the upper limit. To identify the optimal price, a company must thoroughly consider competitors' pricing along with various external and internal factors.

Enterprises could potentially determine prices by embracing a comprehensive pricing approach that factors in different elements.

1. **Cost-Based Pricing**

The highly effective strategy of cost-based pricing ensures that the selling price is determined by expertly applying the desired mark-up to the permissible cost. This approach not only maintains well-regulated profit margins, but also guarantees that pricing policies align with customer perceived value and market rates set by competitors. Consequently, the company's market share remains unscathed and unaffected (Nubbemeyer, 2010; Altomonte et al., 2015).

To streamline the process, we can rely on the ever-reliable cost-plus pricing method, where a standard mark-up is added to the product's cost. This approach is particularly favoured by food and beverage managers when it comes to setting wine prices. For instance, a delightful bottle of wine with a $14 cost can be effortlessly sold for $28, reflecting an astounding 100 percent mark-up on the cost. The profit obtained through this method is an immensely satisfying $14.

2. **Break-Even Analysis and Target-Profit Pricing**

The selling price is determined in cost-based pricing by carefully applying the desired markup to the allowable cost. This approach not only maintains consistent profit margins but also ensures that pricing strategies align with the perceived value held by customers and the market prices established by competitors. As a result, the company's market share remains stable and unharmed (Nubbemeyer, 2010; Altomonte et al., 2015).

To make things simpler, we will use the reliable cost-plus pricing method. This method includes adding a standard markup to the cost of the product. This method is highly favoured by food and beverage managers, especially when setting wine prices. For example, a fantastic bottle of wine that is priced at $14 could potentially be sold for $28, resulting in a remarkably high 100% markup on the cost. The total gross profit from this method is a satisfying $14.

The company must sell a minimum of 30,000 units at a price of $20 in order to achieve a break-even point. At this sales level, the total revenue precisely equals the total cost of $600,000. In order to achieve a target profit of $200,000, the company must exceed expectations by selling at least 50,000 meals, which is equivalent to 137 meals per day. This level of sales would generate a significant revenue of $1 million, covering costs of $800,000 and securing an additional $200,000 for the target profit. If the company chooses a higher price of $25 per meal, it will only require selling 33.34 meals, or 92 meals per day, to reach the desired profit. It is crucial to understand that a higher price lowers the break-even point. The gross profit or contribution from each sale, which aids in covering fixed costs, is determined by the difference between the selling price and the variable cost.

3. **Value-Based Pricing**

Companies are increasingly adopting a strategic approach to pricing, placing a strong focus on the perceived value of the product. The pricing strategy centres around buyers' perceptions of value as the primary factor in setting prices, rather than solely relying on the seller's cost. Value-based pricing, unlike traditional methods, requires the marketer to consider not only the price but also other elements of the marketing mix when finalising the marketing programme. By utilising non-price variables within the marketing mix, the company aims to establish a compelling sense of perceived value in the minds of buyers, ultimately aligning the selling price with this perceived value.

Consider the varying prices set by different restaurants for identical items. An astute individual seeking a delectable cup of coffee and a scrumptious slice of apple pie will discover that they have the opportunity to relish this delightful combination at various establishments, each providing a unique experience. At a drugstore counter, one can indulge in this pleasure for a mere $2, while a family restaurant offers this delightful pairing for a slightly higher price of $3. For those seeking a more refined setting, a hotel coffee shop provides the ambiance and service for $4, while the indulgence of hotel room service can be savoured for $7. Finally, for the epitome of elegance and culinary excellence, an exquisite restaurant awaits, where this delectable combination can be enjoyed for a price of $8. Each subsequent establishment is able to command a higher price due to the value that is added by their distinct service offerings.

To truly embrace the concept of value-based pricing and achieve optimal outcomes, it is crucial that every aspect of a company's operations, including research and development, sales, operations, and marketing, shift their focus from solely considering the product or service, its technical aspects, and cost, to prioritising the customers themselves. This entails understanding their unique needs, preferences, desires, and ultimately, their perception of value. By aligning all these functions towards the customer-centric approach, a company can successfully implement a pricing strategy that is rooted in providing exceptional value to its customers.

Any company utilising the strategy of perceived-value pricing must acquire a profound understanding of the buyers' perception of value in relation to different competitive offers. Occasionally, diligent researchers engage consumers in insightful conversations, questioning them about the monetary worth they associate with each additional benefit incorporated in the offer. One effective means of discerning the amount customers are willing to invest revolves around the utilisation of a technique known as trade-off analysis. In this process, astute researchers inquire about the monetary value buyers would attribute to a hotel room equipped with specific amenities and compare it to the value they would assign to a room without those amenities. Such valuable information grants a valuable insight into which features contribute more value than their associated costs. Should the seller dare to charge an amount exceeding the buyers' perceived value, it will inevitably suffer the consequences of reduced sales. Regrettably, numerous companies fall into the trap of overpricing their offerings, which inevitably leads to underwhelming sales figures. Conversely, some companies make the grave mistake of underpricing their products. While underpriced items may enjoy a remarkable surge in popularity, they fail to generate the desired revenue that could be achieved by elevating the price to match the perceived-value level. Large hotel chains and luxury brands often utilise value-based pricing, a strategy that sets prices according to the guest's perception of the hotel.

4. **Competition-Based Pricing**

A persuasive pricing strategy, called "going-rate pricing," focuses on setting prices primarily based on those of competitors, rather than being solely driven by costs or demand. In this approach, a firm has the freedom to charge the same, more, or less than its major competitors. While some firms may slightly deviate in their pricing, they generally maintain a consistent difference level. For instance, in markets where they compete, a limited-service hotel chain may charge $10 more than a motel. This pricing strategy is highly favoured as it relies on the industry's collective wisdom to determine a price that will generate a fair return. By adhering to the going price, companies also aim to avoid potentially harmful price wars.

Pricing Strategies

1. **New Product Pricing Strategies**

The persuasive pricing strategy called "going-rate pricing" sets prices primarily based on competitors' prices, rather than just considering costs or demand. In this approach, a company can charge the same, more, or less than its major competitors. While some firms may slightly deviate in their pricing, they typically maintain a consistent difference level.

For instance, in markets where they compete, a limited-service hotel chain may establish prices $10 higher than a motel. This pricing method is widely preferred because it relies on the industry's collective wisdom to determine a price that should generate a fair return. By following the going price, companies also aim to avoid harmful price wars, especially in situations where it is difficult to assess price elasticity.

A) **Prestige Pricing**

Hotels and restaurants that aim to establish themselves as luxurious and sophisticated, such as the Royal Penthouse Hotel President Wilson, which is strategically situated near the United Nations Building on the shores of Lake Geneva, have intentionally chosen to enter the market with a premium price. This pricing approach unequivocally emphasises their prestigious status. Similarly, nightclubs may enforce a cover charge to draw in a specific clientele and cultivate an atmosphere of exclusivity. In all of these instances, reducing the price would merely result in a repositioning of the business, failing to attract the intended target market.

B) Market-Skimming Pricing

Price skimming is an incredible strategy to adopt when the market lacks sensitivity towards price. Employing price skimming becomes a logical choice when lowering the price would result in a decrease in revenue. Take, for instance, the scenario where the sole proprietor of a motel in a small town in Louisiana can set exorbitant prices if the demand for rooms surpasses the supply. Similarly, in Araxova, Greece, during the winter season, hotels can confidently set higher prices due to the overwhelming demand for snow activities. The sheer eagerness of consumers to engage in such activities makes them more than willing to pay the premium. Implementing a price skimming strategy is highly effective in the short term. It is important to acknowledge the potential future risks associated with this approach. Competitors will not take notice of the high prices consumers are willing to pay and will choose to enter the market without increasing supply, thus not leading to a subsequent reduction in prices. Price skimming is a commonly employed tactic in industries with substantial research and development expenses, such as pharmaceutical companies and computer firms. In the constantly evolving hospitality industry, it is relatively rare for new competitors to enter the market with ease over an extended period.

According to Chang (2021), the phenomenon of price skimming emerges effortlessly in a state of equilibrium, even without the seller's unwavering dedication, when the buyer possesses an outside option that is contingent upon their specific circumstances. This notion is further supported by Helmold and Helmold (2020).

C) Market-Penetration Pricing

Some companies choose to start with a low initial price instead of a high one in order to swiftly and extensively penetrate the market. This approach is aimed at enticing a large number of buyers and securing a significant market share. For instance, Theodore Zinck's, a cocktail lounge in downtown Dallas, opened with prices that were approximately 20 percent lower than those of their competitors. Zinck's management had successfully negotiated a favourable lease, which gave them a competitive edge. Competitors were unable to match Zinck's lower prices as a result of their higher overhead costs. This pricing strategy enabled Zinck's to rapidly attract a substantial customer base.

Several factors align with adopting a low-price strategy and these include:

- *Market Price Sensitivity*: The market should demonstrate a high sensitivity to pricing changes, ensuring that a lower price stimulates significant market growth.
- *Economies of Scale*: The presence of economies of scale, where costs decrease with higher sales volume, is crucial. This allows for maintaining profitability while offering lower prices.
- *Competitive Barrier*: A lower price should act as a barrier to potential competitors, dissuading their entry due to the challenge of matching the low price while covering their costs.

These considerations make a low-price approach strategically advantageous for businesses seeking to gain a competitive edge and expand their market presence.

2. **Existing-Product Pricing Strategies**

The aforementioned strategies are commonly employed during the launch of a new product. Nevertheless, they can also prove beneficial when applied to existing products. The following strategies are adaptable for use with established products:

A) **Product-Bundle Pricing**

Sellers effectively combine multiple valued products and offer them at a reduced price through product-bundle pricing. Hotels skillfully provide customised weekend packages that include accommodation, meals, and entertainment at a discounted rate. Additionally, they offer business rates that encompass a satisfying breakfast and a complimentary newspaper for a comprehensive guest experience. By adopting the strategy of price bundling, sellers have the capacity to stimulate the purchase of products that consumers may not typically consider acquiring. However, the collective price of the bundle must be positioned at such an appealingly low level that it becomes irresistibly persuasive for customers to seize the opportunity. Furthermore, it is vital that the supplementary items accompanying the core service exhibit a higher perceived value to the customer than the costs incurred for their provision.

Prestigious organisations such as cruise lines, tour wholesalers, and casinos have perfected the practice of product-bundle pricing. For instance, leading cruise lines skillfully present attractive fly-cruise or fly-drive cruise packages that combine the offerings of renowned car rental companies, major airlines, well-established cruise lines, and upscale hotels. These packages are offered at a rate that significantly lowers the cost compared to buying each service individually. The tourist product, with its composite nature, lends itself perfectly to the implementation of product-bundle pricing, a highly effective pricing strategy. An exemplary tourist product consists of a comprehensive package encompassing accommodation, gastronomy, transportation, entertainment, and a myriad of other interconnected services. Given that these components are intended to be experienced collectively rather than in isolation, it becomes imperative to bundle them together and offer them at an accessible and enticing price point.

B) **Psychological Pricing**

Prestigious organisations such as cruise lines, tour wholesalers, and casinos have perfected the practice of product-bundle pricing. For instance, leading cruise lines skillfully present attractive fly-cruise or fly-drive cruise packages that combine the offerings of renowned car rental companies, major airlines, well-established cruise lines, and upscale hotels. These packages are offered at a rate that significantly lowers the cost compared to buying each service individually.

Another element of psychological pricing revolves around reference prices. These are the prices that buyers grasp in their minds and use as a point of reference when examining a particular product. A buyer's reference price can be influenced by current prices, past prices, or the overall buying situation. Reference prices are often associated with popular products. For example, when it concerns a certain kind of restaurant, the majority of customers already possess a preconceived idea about the cost or price range of certain items, like a mug of coffee, a strip steak, or a hamburger. As an example, a pizza chain will confidently promote their medium pizza at a price that they are fully aware is $2 lower than the competition, effectively setting a benchmark price for pizza enthusiasts. Nevertheless, their prices for beverages and additional items will be equal to those of the competition. The benchmark item possesses an extraordinary capability to generate a perception of substantial value, thereby making it inconsequential to decrease the cost of other items.

Customers have a tendency to simplify the complex realm of price information by conveniently disregarding the end figures. It is genuinely captivating to see how the human mind interprets a larger separation between the prices of $0.69 and $0.71, when compared to the relatively insignificant distinction between $0.67 and $0.69. In addition, consumers possess a natural inclination to round figures. A fascinating investigation carried out in a restaurant environment uncovered that individuals have a tendency to round up prices ranging from 0.86 to $1.39 to a whole dollar, whereas prices from $1.40 to $1.79 are commonly rounded up to $1.50. Furthermore, prices ranging from $1.80 to $2.49 are conveniently rounded to $2. Considering this extraordinary behaviour, it is conceivable to propose that a mere price increase of $0.30, from $1.45 to $1.75, might not have a significant effect on the demand. However, it is highly likely that there will be a significant increase in demand when the price reaches $2.05, after surpassing the $1.75 mark.

The magnitude of the price range should be given careful thought, whether it's the shift from $0.99 to $1 or from $9.99 to $10. Even though it's just a $0.01 difference, it should be acknowledged as a significant increase. Taco Bell's value prices, for instance, all stayed below $1, featuring only two digits. Some psychology experts contend that each digit carries symbolic and visual characteristics that should be deeply considered when crafting pricing strategies.

C) **Promotional Pricing**

When companies implement promotional pricing strategies, they strategically reduce the prices of their products, even going as far as selling them below the original list price or even under the cost. Promotional pricing tactics come in various forms. For instance, fast-food establishments may deliberately price certain products as loss leaders, enticing customers to visit the store with the expectation that they will also purchase other items with the standard markups. Doughnut shops, for example, may offer a cup of coffee for a mere 25 cents, fully aware that customers will likely indulge in at least one doughnut to complement their beverage. Similarly, Jack-in-the-Box entices customers with the irresistible deal of three hamburgers for just $1, knowing full well that customers will also opt for French fries and a refreshing soft drink to complete their order. Moreover, during periods of reduced activity, hotels may seize the opportunity to boost their business by presenting customers with exclusive promotional rates that are designed to attract their attention and secure their patronage. Instead of simply offering discounted prices, expertly managed hotels are capable of creating extraordinary experiences: a prestigious Valentine's weekend package that includes a luxurious room, a bottle of champagne upon arrival, an exquisite dinner for two, and a delightful breakfast served in the privacy of your room; or a captivating theatre package that includes a comfortable room, tickets to a mesmerizing play, an intimate dinner for two, and a delectable breakfast for two. These enticing promotions provide guests with a compelling reason to choose their hotel, while the combination of these indulgent offerings enhances the overall value for their esteemed customers. By implementing these carefully curated promotions, they are able to cultivate a positive image that resonates with their discerning clientele, in sharp contrast to the potentially negative perception that can arise from mere price discounting.

The gaming industry for instance, has a profound understanding of the product bundling and promotional pricing. Bruce Rowe, the director of gaming information technology development at Promus Companies, the esteemed parent company of Harrah's, confidently asserted, "We are gracefully immersed in the realm of adult entertainment; our primary offering lies within the realm of gambling, complemented by a multitude of indispensable components such as opulent hotels, exceptional entertainment facilities, and delectable restaurants." Harrah's perceives hotel rooms as a strategic tool to enthrall and empower their esteemed clientele to engage in the exhilarating world of gambling. Rowe further accentuated this notion by emphasizing, "Our esteemed patrons are cordially invited to experience the epitome of luxury within our hotel premises," adding, "However, in order to maximize our revenue, casinos must ensure that these luxurious rooms are always readily available, exclusively for our most esteemed gaming enthusiasts." The pricing strategy employed by Harrah''s for their prestigious hotel rooms aptly reflects the fundamental reality that their core product offering lies within the domain of gaming, with the hotel room serving as merely a supplementary element that further enhances the overall gaming experience.

D) **Value Pricing**

The idea of value pricing can be confusing, but it is vital to grasp its importance. At whatever point a purchaser decides to buy a product or service, irrespective of the price, they must perceive value in that particular offering.

Value pricing has now become synonymous with the term everyday low prices (EDLP), which has been utilised as a powerful marketing strategy by notable players in the hospitality industry, such as Taco Bell and Southwest Airlines.

One must acknowledge that value pricing carries a certain degree of risk. However, when properly conceived and executed, it has the potential to yield neutral outcomes. On the contrary, if not implemented with caution, it can result in catastrophic consequences. At its core, value pricing entails offering a price below that of competitors on a permanent basis. It is important to distinguish this approach from promotional pricing, where prices may be temporarily reduced during special promotions.

Summary

In this chapter, we have thoroughly examined the crucial internal factors that greatly influence pricing decisions. These factors, namely marketing objectives, marketing-mix strategy, costs, and organisational considerations, play a pivotal role in determining the most appropriate pricing strategy. Additionally, we have meticulously identified and defined the external factors that have a significant impact on pricing decisions. These factors encompass the effects of the market and demand, competition, and various other environmental elements. Moving forward, we delved into an in-depth discussion of the general pricing approaches. We made a clear distinction between cost-plus, target profit pricing, value-based pricing, and going-rate pricing. Lastly, we explored the intriguing world of new product pricing strategies, specifically market-skimming pricing and market-penetration pricing, and the strategies related to the existing products.

Activity

1. Define price and state its relevance in achieving organisational objectives.
2. Explain the internal and external factors a manager at the hospitality institution should consider when making pricing decisions.
3. Discuss the general pricing approaches a manager in a hospitality institution can adopt. State two advantages and disadvantages of each approach.

References

Altomonte, C., Barattieri, A., & Basu, S. (2015). Average-cost pricing: Some evidence and implications. *European Economic Review*, *79*, 281–296.

Chang, D. (2021). Optimal sales mechanism with outside options. *Journal of Economic Theory*, *195*, 105279.

Dutta, S., Zbaracki, M. J., & Bergen, M. (2003). Pricing process as a capability: A resource-based perspective. *Strategic Management Journal*, *24*(7), 615–630.

Gibbs, C., Guttentag, D., Gretzel, U., Yao, L., & Morton, J. (2018). Use of dynamic pricing strategies by Airbnb hosts. *International Journal of Contemporary Hospitality Management*, *30*(1), 2–20.

Helmold, M., & Helmold, M. (2020). *Total revenue management (trm)* (pp. 1–12). Springer International Publishing.

Johansson, C., Bras, J., Mondragon, I., Nechita, P., Plackett, D., Simon, P., ... & Aucejo, S. (2012). Renewable fibers and bio-based materials for packaging applications–a review of recent developments. *BioResources*, *7*(2), 2506–2552.

Lockyer, T. (2005). The perceived importance of price as one hotel selection dimension. *Tourism Management*, *26*(4), 529–537.

Mitra, S. K. (2020). An analysis of asymmetry in dynamic pricing of hospitality industry. *International Journal of Hospitality Management*, *89*, 102406.

Nubbemeyer, E. (2010). *A reconsideration of full-cost pricing* (Doctoral dissertation, lmu).

Chapter 16

Distribution Channels

Chapter Outline

- Introduction
- Nature and Importance of Distribution Systems
- Nature of Distribution Channels
- Why Are Marketing Intermediaries Used?
- Distribution Channel Functions
- Number of Channel Levels
- Marketing Intermediaries
- Channel Behaviour and the Organisation
- Channel Behaviour

Objectives

After reading this chapter, you should be able to:

- Describe the nature of distribution channels and explain why marketing intermediaries are used.
- Describe the different marketing intermediaries available to the hospitality industry and the benefits each of these intermediaries offers.
- Discuss channel behaviour and organisation, explaining corporate, contractual, and vertical marketing systems, including franchising.
- Identify factors to consider when choosing a business location.

DOI: 10.4324/9781032688497-16

Introduction

Competition, a global marketplace, electronic distribution techniques, and a perishable product have increased the importance of distribution. Innovative ways of approaching new and existing markets are needed. In this last chapter of unit 3, we are going to study distribution channels as part of the hospitality and tourism marketing mix.

Nature and Importance of Distribution Systems

If we view properties as the heart of a hotel company, distribution systems can be viewed as the company's circulatory system. Distribution systems provide a steady flow of customers. A well-managed distribution system can make the difference between a market-share leader and a company struggling for survival. Many hospitality companies are making greater use of the marketing channels available to them.

Nature of Distribution Channels

A distribution channel is a set of independent organisations involved in the process of making a product or service available to the consumer or business user. The development of a distribution system starts with the selection of channel members. Once members have been selected, the focus shifts to managing the channel. Distribution networks in the hospitality industry consist of contractual agreements and loosely organised alliances between independent organizations. Marketing, distribution systems are traditionally used to move goods (tangible products) from the manufacturer to the consumer. In the hospitality and travel industries, distribution systems are used to move the customer to the product: the hotel, restaurant cruise ship, or airplane.

We first look briefly at traditional distribution systems. These systems provide the framework for the development of hospitality distribution networks. The products used by hospitality and travel companies come through distribution channels; thus, it is important to understand their structure. Graduates of hospitality and tourism programmes often work for companies that distribute products. Graduates with restaurant experience may find themselves working for a company that distributes food or beverages to restaurants. They may sell food-service equipment or table-top items to restaurants and hotels. They may sell supplies to hotels. Some graduates have taken jobs as food brokers. The food broker works as an agent for the manufacturer, trying to create demand for a product. For example, if Mrs. Smith's pies develop a new no-bake pie for the food-service industry, brokers representing Mrs. Smith's pies would introduce the product to food-service managers they think will be interested in using it. The hospitality and travel industries use billions of dollars' worth of products, all moved through distribution channels. These distribution channels create thousands of jobs. The travel industry also provides opportunities for people to serve as intermediaries in the distribution of the tourist product. The tour operator, travel agent, and other specialty chanellers play an important role in linking tourists to destinations.

Why Are Marketing Intermediaries Used?

Why does Shenago China sell its chinaware to restaurants through an intermediary? Although doing so means giving up control over pricing the products, Shenago does gain advantages from selling through an intermediary. The company does not have to maintain several display rooms and a large sales force in every major city. Instead, a restaurant supply company displays, promotes, and makes personal sales calls. The restaurant supply house sells hundreds of other items. Their large assortment makes them a convenient supplier to the restaurant industry. The sales potential from their product assortment allows them to make personal sales calls, send catalogues, and provide other support for the products they represent. Selling through wholesalers and retailers usually is much more efficient than direct sales.

A restaurant manager can make one call to a restaurant supply house and order a French knife, a dozen places, a case of candles, a dozen oyster forks, a case of wine glasses, and a case of cocktail napkins. Each of these items is produced by a different manufacturer, but they are all available through one phone call. To the purchaser, this means access to small quantities of products, because these become part of a large order. This reduces inventory requirements, number of deliveries, and number of processed invoices. Without distribution systems, the restaurateur would have to call individual manufacturers, such as a knife manufacturer, a China company, and a paper company. Each of these manufacturers would receive thousands of calls from individual restaurants. This would create unnecessary work and shipping costs for both the manufacturer and the customer. The restaurants or customers call one distributor and get all of their supplies. The manufacturers can reach many restaurants through one distributor.

Distribution Channel Functions

A distribution channel moves goods from producers to consumers. It overcomes the major time, place, and possession gaps that separate goods and services from those who would use them. Members of the marketing channel perform many key functions:

1. *Information*: gathering and distributing marketing research and intelligence information about the marketing environment.
2. *Promotion*: developing and spreading persuasive communications about an offer.
3. *Contact*: finding and communicating with prospective buyers.
4. *Matching*: shaping and fitting the offer to the buyer's needs, including activities such as manufacturing, grading, assembling, and packaging.
5. *Negotiation*: agreeing on price and other terms of the offer so that ownership or possession can be transferred.
6. *Physical distribution*: transporting and storing goods.
7. *Financing*: acquiring and using funds to cover the costs of channel work.
8. *Risk raking*: assuming financial risks such as the inability to sell inventory at full margin.

The first five functions help to complete transactions. The last three help to fulfil the completed transactions. However, some of these functions do not apply to travel intermediaries. For instance, travel intermediaries do not physically distribute the travel product and do not hold stock.

All these functions have three things in common: they use scarce resources; they can often be performed better through specialisation; they can be shifted among channel members. Shifting functions to the intermediary may keep producer costs and prices low, but intermediaries add a charge to cover the cost of their work. To maintain cost efficiency, it is advisable to assign tasks to channel members who can perform them most effectively. For instance, airlines often promote the use of travel agents. These agents address passenger inquiries, issue tickets, handle payments, and, if there are changes to the passenger's plans, they reissue the ticket. Travel agents are also conveniently located, and, many will deliver a ticket to their clients the same day it is booked. It would not be economically feasible for an airline to set up a similar distribution system.

Number of Channel Levels

Distribution channels can be described by the number of channel levels. Each layer that performs some work in bringing the product and its ownership closer to the final buyer is a channel level. Because the producer and the final consumer both perform some work, they are part of every channel. We use the number of intermediary levels to show the length of a channel.

Channel 1, called a direct marketing channel, has no intermediary level. It consists of a manufacturer selling directly to consumers. For example, a restaurateur may buy produce directly from the grower at a farmer's market. From Figure 16.1, there is a direct marketing channel when travellers bypass the distribution channel and buy directly from suppliers. This has resulted in disintermediation as many destinations and service providers allow potential travellers to book their flights, accomodation and vacations online.

Channel 2 contains one level. In consumer markets, this level is typically a retailer. The Fisherman's Pier restaurant in Geelong, near Melbourne, Australia, purchases its fish from a fisherman's co-op. The co-op markets the fish, allowing the fishers to specialize in fishing, not marketing.

Many of the agricultural products purchased by the hospitality industry come from cooperatives. In the United States, Sunkist, Diamond Walnuts, and Land o' Lake's butter are all producer cooperatives. New Zealand Milk Products Company is also a cooperative and sells dried milk and cheese throughout Southeast Asia and Latin America.

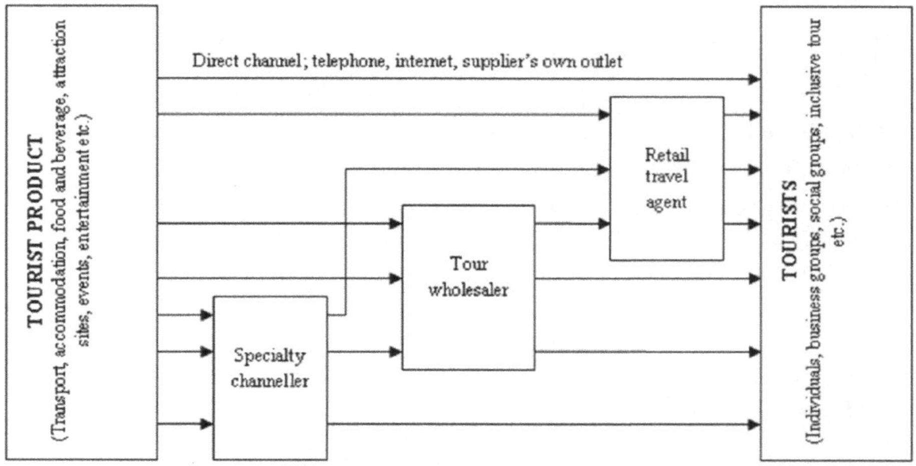

Figure 16.1 Structure of the tourism distribution channel.

Channel 3 contains two levels. In consumer markets, these are typically a wholesaler and a retailer. This type of channel is used by smaller manufacturers.

All the institutions in the channel are connected by several types of flows. These include the physical flow of products, the flow of ownership, payment flow, information flow, and promotion flow. These flows can make channels with only one or a few intermediaries very complex.

Marketing Intermediaries

Many specialized channels are available to hospitality and travel organizations. Components of a hospitality or travel distribution system may include travel agents; wholesalers; specialists; hotel representatives; national, slate, and local tourist agencies; consortia and. reservation systems; global distribution systems; the Internet; and concierges. A manager must choose the intermediaries that will make up the distribution system and the number of levels that the distribution system will have.

Kracht & Wang (2010) indicates that the advance of information and communication technology has not reduced the number of intermediaries in the distribution channel, but rather resulted in an increasingly complex array of intermediaries. The structure of the tourism industry has taken the form of a complex global network, Chou et al. (2020).

Channel Behaviour and the Organisation

Distribution channels are more than simple collections of firms tied together by various flows, they are complex behavioural systems in which people and companies interact to accomplish goals. Some channel systems consist of formal interactions among loosely organized firms. Others consist of formal interactions guided by strong organisational structures. Channel systems do not stand still. New types surface and new channel systems evolve. We now look at channel behaviour and how members organise to do the work of the channel.

Channel Behaviour

A distribution system consists of dissimilar firms that have banded together for their common good. Each channel member is dependent on the others, playing a role in the channel and specialising in performing one or more functions.

Ideally, because the success of individual channel members depends on general channel success, all channel firms should work together. They should understand and accept their roles, coordinate their goals and activities, and cooperate to attain overall channel goals. By cooperating they can more effectively understand and serve the target market.

But individual channel members rarely take such a broad view. They are usually more concerned with their own short-run goals and their dealings with the firms operating closest to them in the channel. Cooperating to achieve overall channel goals sometimes means giving up individual company goals. Although channel members are dependent on each another, they often act alone in their own short-run best interests. They frequently disagree on the roles each should play or who should do what for which rewards. Such disagreements over goals and roles generate channel conflict.

Horizontal conflict is a conflict between firms at the same level of the channel. For example, some Pizza Inn franchisees may complain about other Pizza Inn franchisees cheating on ingredients and giving poor service, thereby hurting the overall Pizza Inn image.

Vertical conflict, which is more common, refers to conflicts between different levels of the same channel.

There could be conflict in the channel, it should however be managed so it does not become unhealthy competition. This conflict can make channel members more active and innovative. But sometimes conflict can damage the channel. For the channel as a whole to perform well, each channel member's role must he specified, and channel conflict must be managed. Cooperation, assignment of roles, and conflict management are attained through strong channel leadership. The channel will perform better if it contains a firm, agency, or mechanism that has the power to assign roles and manage conflict.

Today, the complexity of channels has made it more difficult to manage channel members and act in the best interest of all channel members. Some forms of conflict are the result of management not thinking about how marketing decisions will affect all of a firm's channel members. For example, Embassy Suites had to modify a promotion it developed with Hertz offering cash payments to Hertz customers who were renting cars and staying overnight. The promotion offered Hertz's customers a confirmed hotel reservation a cash voucher if they would switch to an Embassy Suites Hotel. Embassy Suites saw an opportunity to reach hotel customers who were making an immediate purchase, and Hertz saw an opportunity to build a business by offering its customers a cash bonus. It seemed like a good idea for both companies, but the American Society of Travel Agents protested the agreement. They felt that the hotel chain was unfairly taking commissions away from travel agents who had made the original reservations. Both Embassy Suites and Hertz failed to recognize the negative impact that the promotion would have on one of their channel members, the travel agent.

In a large company, the formal organisational structure assigns roles and provides needed leadership. But in a distribution channel made up of independent firms, leadership and power are not formally set. Traditionally, distribution channels have lacked the leadership needed to assign roles and manage conflict. In recent years, new types of channel organizations have appeared to provide stronger leadership and improved performance.

Currently tourists are moving in a new context which has transformed tourism-brokering and changed the way tourists consume. Today's technologically informed client tends to look for experiences and has skills using devices which enable them to obtain information through many different channels. We now have travellers who use technology to find an e-destination, or interact with it and create their own experiences (Sánchez et al., 2020).

Summary

In this chapter, we discussed the nature of distribution channels, and the reasons marketing intermediaries are used. We also gave an exposition on the different marketing intermediaries available to the hospitality industry and the benefits each of these intermediaries offers. We finally discussed channel behaviour and how it can be managed.

Activity

1. Define distribution channels and provide two reasons why effective distribution channels are important in today's competitive hospitality industry.
2. Discuss the three levels of distribution channel options available for hospitality firm management to choose from.

3. Describe the nature of distribution channels, channel behaviour, and how channel conflicts can be managed.

References

Chou, S. F., Horng, J. S., Liu, C. H. S., & Lin, J. Y. (2020). Identifying the critical factors of customer behavior: An integration perspective of marketing strategy and components of attitudes. *Journal of Retailing and Consumer Services, 55*, 102113.

Kracht, J., & Wang, Y. (2010). Examining the tourism distribution channel: Evolution and transformation. *International Journal of Contemporary Hospitality Management, 22*(5), 736–737.

Sánchez-Teba, E. M., García-Mestanza, J., & Rodríguez-Fernández, M. (2020). The application of the inbound marketing strategy on costa del sol planning & tourism board. Lessons for post-covid-19 revival. *Sustainability, 12*(23), 9926.

Chapter 17

Promoting Products: Communication and Promotion Policy and Advertising

Chapter Outline

- Introduction
- The Communication Process
- Determining the Communication Objective
- Design the Message
- Selecting Communication Channels
- Measure the Communications' Results
- Managing and Coordinating Integrated Marketing Communications
- Advertising
- Personal Selling
- Sales Promotion
- Measuring the Communication Effect
- Measuring the Sales Effect

DOI: 10.4324/9781032688497-17

Objectives

After reading this chapter, you should be able to:

- Outline the steps in developing effective communications.
- Describe the ways of setting a total promotional budget.
- Explain each promotional tool and the factors to consider in setting the promotion mix.
- Describe the major decisions in advertising, including setting objectives and budget; creating the advertising message; selecting advertising media; choosing media types, vehicles, and timing, and evaluating advertising.

Introduction

Modern marketing calls for more than developing a good product, pricing it attractively, and making it available to target customers. Companies must also communicate continuously with their present and potential customers. Every company is inevitably cast into the role of communicator and promoter. In this chapter, therefore, we are going to examine communication and promotion policy and advertising as part of developing the hospitality and tourism marketing mix.

What is communicated should not be left to chance. To communicate effectively, companies often hire advertising agencies to develop effective ads, sales promotion specialists to design sales-incentive programmes, and public relations firms to develop corporate images. Salespeople are trained to be friendly, helpful, and persuasive. For any company, the question is not whether to communicate, but how much to spend and in what ways.

Happ and Ivancsó-Horváth (2018) believe that tourism has entered into a new era of digital tourism. In many countries, the strategy of digital tourism has already been defined. People's interest in digital techniques and their use has changed over the last few decades and this change has to be followed in tourism as well. Without digitisation, tourism will not be competitive either with foreign destinations or with service providers.

The Communication Process

Marketing communication comprises the entirety of the activities undertaken to create and market an idea of a company and its product, and also distributing the idea among various groups of people. Communication is a highly dynamic phenomenon, and we can observe the continuous development of new forms of marketing communication: internet communication, mobile communication, viral marketing, buzz marketing, guerrilla marketing, advergaming, and ambush marketing (Kuczamer-Klopotowska 2017).

Today there is a new view of communications as an interactive dialogue between the company and its customers that takes place during the preselling, selling, consuming, and post-consuming stages. Companies must ask not only "How can we reach our customers?" but also "How can our customers reach us?"

Thanks to technological breakthroughs, people can now communicate through traditional media (newspapers, radio, telephone, television), and through newer media forms (computers, fax machines, cellular phones, and pagers). By decreasing communication costs, the new technologies have encouraged more companies to move from mass communications to more targeted communication and one-to-one dialogue.

The salesperson's manner and dress, the place's decor, the company's stationery—all communicate something to the buyers. Every brand contact delivers an impression that can strengthen or weaken a customer's view of the company. The whole marketing mix must be integrated to deliver consistent messages and strategic positioning.

A company's total marketing communications programme, called its promotion mix, consists of a specific blend of advertising, sales promotion, public relations, and personal selling to achieve advertising and marketing objectives. The four major promotion tools are defined next:

1. *Advertising*: any paid form of non-personal presentation and promotion of ideas, goods, or services by an identified sponsor.
2. *Sales Promotion*: short-term incentives to encourage the purchase or sales of a product or service.
3. *Public Relations*: building good relations with the company's various publics by obtaining favourable publicity, developing a good corporate image, and handling or heading off unfavourable rumours, stories, and events.
4. *Personal Selling*: Oral presentation in a conversation with one more prospective purchaser to make sales.

The starting point is an audit of all the potential interactions target customers may have with the company. For example, someone interested in planning a vacation would talk to others, see television ads, read articles, talk to a travel agent, and look for information on the Internet. The marketer needs to assess which experiences and impressions will have the most influence at each stage of the buying process. This understanding will help marketers allocate their communication dollars more efficiently.

There are five steps in developing effective communication. The marketing communicator must

(1) Identify the target audience.
(2) Determine the communication objectives.
(3) Design the message.
(4) Select the communication channels.
(5) Measure the communications results process.

Identifying the Target Audience

A marketing communicator starts with a clear target audience in mind. The audience may be potential buyers or current users, those who make the buying decision, or those who influence it. The audience may be individuals, groups, special publics, or the general public. The target audience will heavily affect the communicator's decision on what will be said, how it will be said, when it will be said, where it will be said, and who will say it. To create effective communication, a marketer must understand the target audience by creating a message that will be meaningful to them in a medium they will understand. For example, a study of bed and breakfast owners found that other than word of mouth, they felt the two most important communication channels were brochures and guidebooks. Managers need to understand their target markets before they can communicate with them.

Determining the Communication Objective

Once a target audience has been defined, the marketing communicator must decide what response is sought. Of course, in most cases the final response is purchase. But purchase is the result of a long process of consumer decision-making. The marketing communicator needs to know where the target audience stands in relation to the product and to what state it needs to be moved.

The Indian tribes of South Dakota wished to increase significantly tourist visitation to their reservations. Their objectives were:

1. To provide guests for bed and breakfast operations.
2. To increase the market for Indian products.
3. To participate in other tourism-related incomes.
4. To correct misconceptions about the American Indian.
5. It was deemed important to show that the Lakota, Dakota, and Nakota people are living cultures.

This combination of economic and cultural education objectives led to the development of the Alliance of Tribal Tourism Advocates (ATTA) as a communication vehicle instead of depending on the South Dakota Department of Tourism or other organisations. Indians would promote themselves. "If you want to visit an Indian, the best person to talk with is a Native American," said Ronald L. Neiss acting director of ATTA and a member of the Rosebud Sioux Tribal Council.

Design the Message

In order to define the desired audience response, the communicator turns to develop an effective message. Ideally, the message should get attention, hold interest, arouse desire, and obtain action (a framework known as the AIDA model). In designing a promotional message for a tourist destination, getting attention will require running a PR campaign among the target market. Holding interest could be achieved through a social media campaign. Desire could be aroused through an advertising campaign in the local media and action could be obtained by offering discounted packages through local tour operators. In practice, few messages take the consumer from awareness to purchase, but the AIDA framework does suggest the desirable qualities of a good message. The need to ensure that all marketing and non-marketing information reflects the brand personality and create a strong message across all its channels has become a critical success factor in the highly competitive market place of tourism and hospitality services (McCabe, 2010).

Selecting Communication Channels

The communicator must now select channels of communication. There are two broad types of communication channels: personal and non-personal.

Personal Communication Channels

In personal communication channels, two or more people communicate directly with each other. They might communicate face to face, person to audience, over the telephone, or even through the mail. Personal communication channels act effectively because they allow for personal addressing and feedback.

Some personal communication channels are controlled directly by the communicator. For example, company salespeople contact buyers in the target market. But other personal communications about the product may reach buyers through channels not controlled directly by the company. These might include independent experts making statements to target buyers, such as consumer advocates and consumer buying guides, or they might be neighbours, friends, family members, and associates talking to target buyers. This last channel, known as word-of-mouth influence, has a considerable effect in many product areas. Personal influence carries great weight for products that are expensive, risky, or highly visible. Hospitality products are often viewed as being risky because they cannot be tried out beforehand. Therefore, personal sources of information are often sought before someone purchases a travel package, selects a restaurant, or stays at a hotel.

Measure the Communications' Results

Alter sending the message, the communicator must evaluate its effect on the target audience. This involves asking the target audience whether they remember the message, how many times they saw it, what points they recall, how they felt about the message, and their past and present attitudes towards the product and company. The communicator would also like to measure behaviour resulting from the message: how many people bought a product, talked to others about it, or visited the store.

Managing and Coordinating Integrated Marketing Communications

The company must now divide the total promotion budget among the major promotional tools: advertising personal selling, sales promotion, and public relations. It must carefully blend the promotion tools into a coordinated promotion mix that will achieve its advertising and marketing objectives.

Companies within the same industry differ greatly in how they design their promotion mixes. Thus, a company can achieve a given sales level with varied mixes of advertising, personal selling, sales promotion, and public relations.

Companies are always looking for ways to improve promotion by replacing one promotion tool with another that will do the same job at less expense. Many companies have replaced a portion of their field sales activities with telephone sales and direct mail. Others have increased their sales promotion spending in relation to advertising to gain quicker sales.

Designing the promotion mix is even more complex when one tool must be used to promote another. Thus, when McDonald's decides to run million-dollar sweepstakes in its fast-food outlets (a sales promotion), it has to run ads to inform the public. Many factors influence the marketer's choice of promotion tools.

Nature of Each Promotion Tool

Each promotion—advertising, personal selling, sales promotion, and public relations—has unique characteristics and costs. Marketers must understand these characteristics to select their tools correctly.

Advertising

Because of the many forms and uses of advertising, generalising its unique qualities as a part of the promotion mix is difficult. Yet several qualities can be noted. Advertising's public nature suggests that the advertised product is standard and legitimate. Because many people see ads for the product, buyers know that purchasing the product will be publicly understood and accepted. Advertising also allows the seller to repeat a message many times. Large-scale advertising by a seller says something positive about the seller's size, popularity, and success.

Advertising can be used to build a long-term image for a product (such as Four Seasons or McDonald's ads) and also stimulate quick sales (as when Embassy Suites in Phoenix advertises a promotion for the Fourth of July holiday). Advertising can reach masses of geographically dispersed buyers at a low cost per exposure.

Advertising also has shortcomings. Although it reaches many people quickly, advertising is impersonal and cannot be as persuasive as a company salesperson. Advertising is able to carry on only one-way communication with the audience, and the audience does not feel that it has to pay attention or respond. In addition, advertising can be very costly. Although some forms, such as newspaper and radio advertising, can be done on small budgets, other forms, such as network TV advertising, require very large budgets.

Personal Selling

Personal selling is the most effective tool at certain stages of the buying process, particularly in building buyer preference, conviction, and purchase. Compared with advertising, personal selling has several unique qualities. It involves personal interaction between two or more people, allowing each to observe the other's needs and characteristics and make quick adjustments. Personal selling also lets all kinds of relationships spring up, ranging from a matter-of-fact selling relationship to a deep personal friendship. The effective salesperson keeps the customer's interests at heart to build a long-term relationship. Finally, with personal selling, the buyer usually feels a greater need to listen and respond, even if the response is a polite "no thank you."

Sales Promotion

Sales promotion includes an assortment of tools, coupons, contests, cents-off deals, premiums, and others, and these tools have many unique qualities. They attract consumer attention and provide information that may lead the consumer to buy the product. Sales promotions offer strong incentives to purchase by providing inducements or contributions that give additional value to consumers, and they invite and reward quick responses. Advertising says, "buy our product"; sales promotion says, "buy it now."

Companies use sales promotion tools to create a stronger and quicker response. Sales promotion can be used to dramatise product offers and to boost sagging slides. Its effects are usually short-lived, however, and are not effective in building long-run brand preference.

Measuring the Communication Effect

Measuring the communication effect reveals whether an ad is communicating well. Called copy testing, this process can be performed before or after an ad is printed or broadcast. There are three major methods of advertising pretesting. The first is direct rating, in which the advertiser exposes a consumer panel to alternative ads and asks them to rate the ads. Direct ratings show how well the ads attract attention and how they affect consumers. Although it is an imperfect measure of an ad's actual impact, a high rating indicates a potentially effective ad. In portfolio tests, consumers view or listen to a portfolio of advertisements, taking as much time as they need. The interviewer then asks the respondent to recall all the ads and their contents. The recall can be either aided or unaided by the interviewer. Recall level indicates the extent to which an ad stands out and how well its message is understood and remembered. Laboratory tests use equipment to measure consumers' physiological reactions to an ad: heartbeat, blood pressure, pupil dilation, and perspiration. The tests measure an ad's attention-getting power but reveal little about its impact on beliefs, attitudes, or intentions.

There are two popular methods of post-testing ads. Using recall tests, the advertiser asks people who have been exposed to magazines or television programmes to recall everything they can about the advertisers and products they saw. Recall scores indicate the ad's power to be noticed and retained. In recognition tests, the researcher asks readers of, for instance, a given issue of a magazine to point out what they have seen. Recognition scores can be used to assess the ad's impact in different market segments and to compare the company's ads with those of competitors.

Measuring the Sales Effect

What quantity of sales are caused by an ad that increases brand awareness by 20 percent and brand preference by 10 percent? The sales effect of advertising is often harder to measure than the communication effect. Sales are affected by many factors besides advertising, such as product features, price, and availability. One way to measure sales effect is to compare past sales with past advertising expenditures. Another is through experiments.

To spend a large advertising budget wisely, advertisers must define their advertising objectives, develop a sound budget, create a good message, make media decisions, and evaluate the results.

Advertising draws much public attention because of its power to affect lifestyles and opinions. Advertising faces increased regulation to ensure that it performs responsibly.

Summary

In this chapter, we outlined the steps in developing effective communications. We also defined the ways of setting a total promotional budget: affordable, percentage-of-sales, competitive-parity, and objective and task methods. We then explained each promotional tool—advertising, personal selling, sales promotion, and public relations—and the factors in setting the promotion mix.

Activity

1. Define marketing communication and outline the steps involved in an effective communications process. Illustrate how these steps can be applied in designing effective communications for a restaurant.
2. Discuss the two major types of communication channels and provide two advantages of using each channel.
3. What is integrated marketing communication? Discuss the nature of each promotional tool (advertising, personal selling, sales promotion, and public relations).
4. Discuss the three major ways of determining the effectiveness of a communication campaign in the hospitality industry.

References

Happ, E., & Ivancsó-Horváth, Z. (2018). Digital tourism is the challenge of future–a new approach to tourism. *Knowledge Horizons. Economics, 10*(2), 9–16.

Kuczamer-Kłopotowska, S. (2017). Sensory marketing as a new tool of supporting the marketing communication process in tourism services sector. *Handel Wewnętrzny, 367*(2), 226–235.

McCabe, S. (2010). *Marketing communications in tourism and hospitality*. Routledge.

Chapter 18

Promoting Products: Public Relations and Sales Promotion

Chapter Outline

- Introduction
- Public Relations
- Major Activities of Public Relations Departments
- Press Relations
- Product Publicity
- Corporate Communication
- Lobbying
- Counselling
- Publicity
- Sales Promotion
- Setting Sales Promotion Objectives
- Selecting Sales Promotion Tools
- Consumer Promotion Tools

Objectives

After reading this chapter, you should be able to:

- Understand the concept of public relations and its relevance to the H&T industry.

DOI: 10.4324/9781032688497-18

- Explain the different public relations activities: press relations, product publicity, corporate communications, lobbying, and counselling.
- Explain the public relations process.

Introduction

Welcome to this chapter on public relations and Sales promotion. In this chapter, we are going to discuss the role public relations and sales promotion tools play in the promotion of hospitality and tourism products.

Public Relations

> "Public relations, perhaps the most misunderstood part of marketing communications, can be the most effective tool." Definitions for public relations differ widely. We think that this definition by Hilton International best fits the hospitality industry: "The process by which we create a positive image and customer preference through third-party endorsement."

Public relations (PR) is an important marketing tool that until recently was treated as a marketing stepchild. PR is moving into an explosive growth stage. Companies are realising that mass marketing is no longer the answer to some of their communication needs. Advertising costs continue to rise. While audience reach continues to decline. Sales promotion costs have also increased as channel intermediaries demand lower prices and better commissions and deals. Personal selling can cost over $500 a call. In this environment, public relations holds the promise of a cost-effective promotional tool. The creative use of news events, publications, social events, community relations, and other PR techniques offers companies a way to distinguish themselves and their products from their competitors.

Public relations emphasise the value of relationships between organisations and their environments. According to one of the most frequently cited definitions, public relations is "the management function that establishes and maintains mutually beneficial relationships between an organization and the publics on whom its success or failure depends" (Dozier and Broom, 2009, p. 25; Sha, 2011).

The public relations department of cruise lines, restaurant chains, airlines, and hotels are typically located at corporate headquarters. Often, its staff is so busy dealing with various publics—stockholders, employees, legislators, and community leaders—that PR support for product-marketing objectives tends to be neglected. Many four- and five-star hotel chains have corrected this deficiency by hiring local public relations managers.

In the past, it was common for the marketing function and PR function to be handled by two different departments within the firm. Today these two functions are increasingly integrated. There are several reasons for this integration. First, companies are calling for more market-oriented PR. They want their PR departments to manage PR activities that contribute towards marketing the company and improving the bottom line. Second, companies are establishing marketing PR groups to support corporate/product promotion and image-making directly. Thus, marketing PR, like financial PR and community PR serves a special constituency, the marketing department.

Major Activities of PR Departments

PR departments perform the five activities discussed below, not all of which feed into direct product support.

Press Relations

Press relations aim to place newsworthy information into the news media to attract attention to a person, product, or service. One reason for the growth of press relations in the hospitality industry is its credibility. Most types of publicity are viewed by the consumer as third-party information. A favourable write-up of a restaurant in the local newspaper by the food editor has more impact than an advertisement written by the restaurant's management.

Product Publicity

Product publicity involves various efforts to publicise specific products. New products; special events, such as food festivals; redesigned products, such as a newly renovated hotel; and products that are popular because of current trends, such as non-fat desserts and leisure travel packages, are all potential candidates for publicity.

Corporate Communication

Corporate communication covers internal and external communications and promotes understanding of the organisation. One important marketing aspect of corporate communication is communication directed towards employees, such as company newsletters. Companies also need to manage their communication with their stockholders to make sure the stockholders understand the company's goals and objectives.

Lobbying

Lobbying involves dealing with legislators and government officials to promote or defeat legislation and regulation. Large companies employ their lobbyists, whereas smaller companies lobby through their local trade associations.

Counselling

Counselling involves advising the management about public issues and company positions and image. Counselling is important when there may be sensitive issues associated with the business. For example, water is a scarce commodity in Las Vegas. Major resorts with water displays, such as the Mirage, counsel their managers on the resort's water conservation efforts, such as recycling the hotel's wastewater to be used in the hotel's fountains.

Publicity

Publicity is a direct function of public relations. Publicity is the task of securing editorial space, as opposed to paid space, in print and broadcast media to promote a product or a service. Publicity is a popular PR tool used in the five activities mentioned earlier. Some popular uses of publicity are described next. One of the uses of publicity is to assist in the launch of new products. For example, when the Hard Rock Cafe announced that it was going into the hotel business with the development of the first Hard Rock Hotel, the media covered the event during the initial announcement and the ground-breaking ceremonies. Later, when the hotel opened, a concert staged at the hotel featuring Sheryl Crow was broadcast on MTV. This concert, the uniqueness of the hotel, and a concert the following day by the Eagles and Sheryl Crow ensured that the opening of the hotel received worldwide publicity.

Publicity is also used with special events. To be successful, the press release developed to gain the publicity must be of interest to the target audience of the media the company is targeting; for example, a food editor will be interested in recipes and food history that will have value for her readers.

Sales Promotion

Sales promotion consists of short-term incentives to encourage the purchase or sale of a product or service. Sales promotion includes a variety of promotional tools designed to stimulate earlier or stronger market responses. It includes consumer promotion (samples, coupons, rebates, price-off, premiums, contests, demonstrations), trade promotion–buying allowances (free goods, cooperative advertising, and push money), and sales force promotion (bonuses and contests.) Often a well-planned sales promotion can result in publicity. The Omni San Antonio Hotel offered a Teacher's Appreciation Special in recognition of their contribution as educators. This sales promotion created goodwill among the teachers and the community and generated publicity for the hotel. It also generated room sales during a short period. Applebee's gives a free child's meal to students who make an A. This rewards students who get good grades and provides the school with a no-cost way of recognising students who have done well. In addition to the public relations benefits, it also brings the child's parents and siblings to Applebee's when the free meal is redeemed. Thus, Applebee generates profitable sales from the promotion.

Setting Sales Promotion Objectives

Sales promotion objectives vary widely. Consumer promotions can increase short-term sales or they can be used to help build long-term market share. The objective may be to entice consumers to try a new product, lure consumers away from competitors, or hold and reward loyal customers. For the sales force, objectives include building stronger customer relations and obtaining new accounts.

Sales promotions should be designed to enhance consumer franchise; that is, they should promote the product's positioning and include a sales message. Ideally, the objective is to build long-run consumer demand rather than to prompt temporary brand switching. If properly designed, every sales promotion tool has a consumer franchise-building potential.

Selecting Sales Promotion Tools

Many tools can be used to accomplish sales promotion objectives. The promotion planner should consider the type of market, the sales promotion objectives, the competition, and the costs and effectiveness of each tool. The main consumer promotion tools are described next.

In the tourism industry, promotions can play a critical role for creating destination awareness, unique identity, image, and loyalty. The tourism promotion mix consists of the activities of advertising, sales promotion, direct marketing, events and experience marketing, online marketing, and so on that a destination marketer uses to communicate its information, features and benefits to target customers (Pramanik et al., 2020).

Consumer Promotion Tools

The main consumer promotion tools include samples, coupons, premiums, patronage rewards, point-of-purchase displays, contests, sweepstakes, and games.

Samples: Samples are offers of a trial amount of a product. Some samples are free. For others, the company charge; a small amount to offset its cost, McDonald's offered a cup of coffee and an apple-bran muffin for $1. Normally, the coffee and the muffin were offered for 95 cents each. The promotion was designed to get customers to try the muffin. Some people do not eat bran muffins and by "charging" 5 cents for the muffin, McDonald's avoided giving the muffin away to customers who would never buy one in the future.

The Inn on the Park in Houston invited potential customers and influential community members to stay in the luxury hotel at no charge. The promotion accomplished two objectives:

(1) Salespeople were aided in selling corporate contracts because many of their potential customers had experienced the hotel; and
(2) Positive word of mouth about the hotel was created. Sampling is the most effective but also the most expensive way to introduce a new product.

Sampling by the staff who are employed by a hospitality firm such as a hotel, restaurant, or ski resort can be a very useful educational and promotional device. Thorough knowledge of the product is particularly beneficial in up-selling. It is difficult for anyone to recommend a premium-priced Bordeaux or California Merlot if they have no idea how the wine tastes. The sales and reservation staff of a hotel or resort can more convincingly sell a prospect on the idea of upgrading to a poolside cabana or suite if they have personal knowledge of the product.

How does the staff obtain personal knowledge of the product or services of a company? Several successful approaches have been used to accomplish staff product knowledge.

1. Provide continuous training programmes. Invite suppliers such as vintners, cheese producers, and gourmet coffee distributors to provide samples and assist with product training.
2. Offer sales and performance incentives that include prizes on the property, such as a five-course meal, a month's use of the health club, or a weekend in the deluxe suite.
3. Create an employee's day in which the staff has full use of the facility. Country clubs often provide a special day in which employees and sometimes their families are treated to exclusive use of the pool, the golf course, the restaurant, and even the ballroom for an evening dance.
4. Share product information with employees through newsletters or product brochures. Often, product information brochures remain only in the offices of the purchasing department, the Food & Beverage manager, or some other executive office.
5. Talk continuously about the company's products and services in an upbeat manner, People tend to forget the many positive attributes of the facilities and the services that surround us daily.

Preston L. Smith, the president and CEO of Ski Limited, regularly sent memos to company managers urging them to hit the slopes. Smith personally manages to ski over sixty times each season. "Everyone skis here. It's a way of sharing the customer's experience. It's also a way to achieve personal growth because skiing is exhilarating and exciting."

Coupons: Coupons are certificates that offer buyers savings when they purchase specified products. More than 220 billion coupons are distributed in the United States each year, with a total face value of more than $55 billion. Coupons can be mailed, included with other products, or placed in ads. Coupons are most popular in the restaurant industry; however, hotels, rental car companies, tourist attractions, and cruise lines also use coupons.

American Express cardholders received coupon packs featuring mid- and upscale restaurants. The prestige of American Express allows these restaurants to use coupons without detracting from their image.

Many professional marketing consultants and observers of marketing and sales practices feel that too much promotion creates a commodity out of a differentiated product. It is argued that companies spend millions of dollars and years of effort to develop a distinct image and a high level of product differentiation in the minds of consumers, only to have it destroyed by promotions.

Packages: Promotions often involve packages of a number of the company's products. Packages are particularly popular with hotels and resorts that have several products to offer. The Ritz Carlton in Tyson's Corners developed a Fine Art of Cuisine weekend. The weekend features gourmet meals matched with the appropriate wines. The hotel developed packages around the meals, tastings, and demonstrations. For $600 guests receive a room, tickets to the Grand Wine Tasting, and the Chefs brunch. Packages can also be developed around local events. The Best Western Palm Beach Florida created a three-night package that includes a room for three nights, tickets to two baseball games (several teams had spring training near the hotel), a continental breakfast, and an evening cruise on a casino ship. The price of the package is $256.27 Promotions such as these bring in business during a slow period and create a memorable experience for the guest.

Premiums: Premiums are goods offered either free or at low cost as an incentive to buy a product. For example, fast-food restaurants often offer a free promotional glass instead of their normal paper cup. A self-liquidating premium is a premium sold to consumers who request it. McDonald's in Australia offered Batman figures for 95 cents with the purchase of a burger.

Point-of-Purchase Displays: Point-of-purchase (POP) promotions include displays and demonstrations that take place at the point of purchased or sale. For example, a representative of Richmond Estate Wines might offer a taste of their wines in the Robina Tavern package store.

The value of POP has long been recognised by the retailing industry and is making rapid inroads in restaurants, hotels, auto rental companies, and other hospitality industry firms. Hospitality firms have discovered that POP may be used.

(1) To disseminate information about the company's products or services, and
(2) To sell additional products and services, thus adding to gross revenue.

Hotels use display racks in the lobby to promote other hotels in the chain and additional services, from valet parking to sleigh rides. Restaurants such as Perkins, the Village Inn, and Denny's use the space near the cash register to create eye-catching displays of bakery items and desserts to be taken home by the guests.

Summary

In this chapter, we have learnt about the different public relations activities: press relations, product publicity, corporate communications, lobbying, and counselling. We also examined sales promotion and some sales promotion activities.

Activity

1. Define hospitality public relations and provide two importance of public relations to the hospitality industry.
2. Describe the public relations function and the role it plays in promoting the services of a hospitality firm.
3. Discuss the various public relation tools and indicate how each tool can be used in the H&T industry.
4. Define publicity and state how different it is from public relations.
5. Define sales promotion and discuss the role of public relations and publicity in the success of sales promotion in the hospitality industry.
6. Discuss the various sales promotional tools and how managers in the hospitality industry could use them effectively to achieve their promotional objectives.

References

Dozier, D. M., & Broom, G. M. (2009). The centrality of practitioner roles to public relations theory. In D. M. Dozier & G. M. Broom (Eds.), *Public relations theory ii* (pp. 137–170). New York, NY: Routledge.

Pramanik, S. A. K., & Rakib, M. R. H. K. (2020). Conceptual analysis on tourism product and service promotion with special reference to Bangladesh. In *Tourism marketing in Bangladesh* (pp. 109–126). Routledge.

Sha, B. L. (2011). 2010 practice analysis: Professional competencies and work categories in public relations today. *Public Relations Review, 37*(3), 187–196.

Chapter 19

Promoting Products: Drawing the Marketing Communications Plan

Chapter Outline

- Introduction
- Developing a Marketing Communications Plan
- Budget and Schedule
- Establishing the Total Marketing Communications Budget
- Affordable Method
- Percentage of Sales Method
- Competitive Parity Method
- Objective and Task Method

Objectives

By the end of this chapter, the student should be able to:

- Understand the importance of developing a marketing communication plan.
- Draw a marketing communication plan.
- Prepare a marketing communication budget.

DOI: 10.4324/9781032688497-19

Introduction

The principal objective of marketing communications is to enhance awareness of an enterprise, its offerings, and its positioning through tools and platforms such as brochures, press releases, websites, and trade show exhibitions. The quintessential duty of a marketing communications manager is to orchestrate an integrated, consistent, and cross-team approach to these endeavours—one that accentuates a firm's narrative to its target demographic and galvanises customers towards a purchase. In this chapter, we shall delve into the intricacies of the marketing communications plan.

To craft an efficacious marketing communications plan, exhaustive research is imperative (Nguyen & Mogaji, 2022). It demands a profound comprehension of your target audience and the multifaceted intricacies of purchasing, selling, and communication dynamics. Once equipped with such insights, you can ascertain your aspirations from your marketing endeavours, discern the pivotal messages for your clientele, and devise optimal strategies to relay that information. Budgetary considerations and scheduling are paramount, and it's crucial to assess any potential restrictions they might impose on your campaign. Ideally, your marketing blueprint should elucidate the communications trajectory systematically. The burgeoning interest in the efficacy of marketing communication, regarded as a linchpin in establishing robust brands, has been highlighted in recent academic pursuits (Keller, 2009; Itani et al., 2023).

Developing a Marketing Communications Plan

Marketing communications (often abbreviated as MarCom) encompass customer-oriented materials such as brochures, press releases, digital platforms, and trade show displays. The intimate nexus of MarCom planning with the consumer renders it as one of the pivotal undertakings for any establishment. Meticulously curated MarCom strategies:

1. Ensure a cohesive brand voice across a myriad of products, geographical locales, and customer touchpoints.
2. Facilitate a diverse consortium of professionals to forge integrated communication tools.
3. Amplify your narratives amidst the cacophony of the marketplace, consistently resonating with the intended recipients.
4. Optimise outcomes against both temporal and financial investments.

Commencing with in-depth research, one then embarks on astute decision-making:

1. *Target Audience Profiles*: Recognise both your extant and prospective clientele. What communication methodologies resonate most with them?
2. *Processes of Buying, Selling, and Communication*: Comprehend your customer's purchase inclinations, evaluate your organisational sales strategies, and identify the communication channels that most effectively bridge these realms.

With an intimate grasp of these dynamics, pivotal decisions await:

1. *Objectives*: What outcomes do you anticipate from your MarCom initiatives? Which customer behaviours are you aiming to foster?
2. *Messages*: What pivotal insights do you wish to impart to your clientele?
3. *Vehicles*: Identify the optimal mediums to articulate your messages.
4. *Budget and Schedule*: Ascertain the financial implications of your chosen mediums, juxtapose this against your budgetary allocations, and strategise their deployment timeline.

Furthermore, the breadth of the plan's focus warrants consideration. MarCom strategies can be tailored at the product echelon, the product suite, or the overarching corporate or brand image, with a potential meld of these tiers.

Target Audience Profiling

Knowing your customer applies to virtually every marketing activity, and it's doubly true for MarCom planning. First, learn about your target audience's specific demographics and psychographics, and then take the time to understand:

- How customers prefer to receive information about your type of offerings.
- How customers use this information during the buying process.
- Buying, selling, and communications processes.
- The primary role of marketing communications is to maximise sales by aligning your selling process with your customer's buying process.

Your customer's buying process is the steps that they follow when making a purchase.

Your selling process is the steps that your sales force goes through when making a sale.

Your communication process creates the customer mindset that you are trying to achieve during each stage of the buying and selling processes.

Case Study: Understanding Your Customer in MarCom Planning

Understanding your customer is the cornerstone of any successful marketing activity, and its importance is magnified when it comes to marketing communication (MarCom) planning. Establish a compelling MarCom strategy for a Ghanaian company, Green Cocoa.

Background of Green Cocoa

Green Cocoa, established in 2018, is a pioneering cocoa processing company in Ghana, blending traditional cocoa cultivation with modern sustainability practices. Founded on eco-friendly operations, community empowerment, and celebrating Ghana's cocoa heritage, Green Cocoa has become one of the leading companies in the region. The company's roots can be traced back to the founders' family-owned farms, which spanned generations. Green Cocoa uses cutting-edge technology to minimise wastage and maximise the retention of beans' natural flavours and nutrients. The company also educates local farmers about sustainable farming practices, creating a positive environmental impact throughout the Ghanaian cocoa industry. A significant portion of Green Cocoa's profits are reinvested in community development projects, ensuring prosperity is shared with those who contribute to its success.

Strategy

1. **Dive Deep into Demographics and Psychographics**

- *Demographics*: These are quantitative aspects of your target audience. Consider factors such as age, gender, income level, education, occupation, and location.

- *Psychographics*: This delves into the qualitative aspects like interests, hobbies, values, and lifestyle choices.

For Green Cocoa, a cocoa processing company in Ghana, demographics might show that most of their customers are women aged 25–40 years, with a university degree, living in urban areas of Accra, Tema, and Kumasi. Psychographics could reveal that these customers value organic products, often indulge in luxury chocolates, and prioritise sustainable farming practices.

2. **Preferred Information Channels**

- Ascertain how your target audience likes to receive information about offerings similar to yours. Is it through social media, emails, in-person events, or other channels?

Green Cocoa finds out through surveys that their target audience primarily discovers new cocoa products through Instagram and food blogs.

3. **Information Utilisation during Purchase**

- Determine how potential customers use the information you provide when they are in the decision-making process. Do they look for reviews, check for sustainability credentials, or compare products side by side?

Green Cocoa customers might first read about the product on Facebook, then check for reviews on Green Cocoa's Instagram page, and finally visit their website to check for certifications before making a purchase.

4. **Aligning Communication, Buying, and Selling Processes**

- *Buying Process*: The journey your customer embarks upon when considering and finally deciding to purchase your product.
- *Selling Process*: The path your sales team or strategy follows to lead a potential customer from awareness to purchase.
- *Communication Process*: The strategy employed to establish a certain perception or mindset in your customer during both buying and selling processes.

When a potential Green Cocoa customer identifies a need for quality cocoa, the communication process should first introduce them to Green Cocoa's premium range (awareness). As they consider their options, they might be offered a limited-time discount (interest and desire). Finally, a seamless online purchase system ensures they make the buy (action). On the selling side, Green Cocoa might initiate contact through targeted Instagram ads (awareness), engage through interactive posts and stories (interest and desire), and finally direct them to their online store with a tempting offer.

To conclude, to maximise sales and ensure effective MarCom, it's essential to align your selling strategy with the buying behaviour of your customers, fine-tuned by the insights gained through understanding their demographics and psychographics.

Effective marketing communication begins with a thorough understanding of these processes. To move a customer from identifying a need to making a purchase, your communications must first make them aware of your offering, then build credibility, and so on. Identifying and understanding the different stages of these processes makes it possible for you to craft marketing messages more and select communications vehicles more. Remember, these are typical processes. Different customers (governments, businesses, consumers) have different buying processes, and your selling process should vary depending on your business model.

Objectives

MarCom plans should build on your marketing strategy plan. However, many companies have no formal statement of high-level marketing strategy, so the MarCom plan must cover more ground. This means creating a number of MarCom objectives that specify what you want customers to do after experiencing your marketing communications. For example, should they visit your website? Call your salesperson? Consider you a top provider in your industry?

It's also important to decide how you measure success. The more quantifiable the metrics—a 10% increase in market share or the generation of 100 new leads—the better.

Messaging

Marketing messages are touchstones that focus and unify your marketing vehicles. Remember, marketing messages:

- About individual products can focus on specific features and benefits.
- About a product line are typically limited to highlighting similarities, differences, and relative positioning among products.
- About your company image typically address overall branding.
- Must be consistent and compatible with each other and your company branding.
- Must reflect the stage of the communications process that you are addressing. Creating awareness requires different messages than creating preference.

While studies argue that a number of factors can influence information search and processing behaviours (Kim, 2014), the way information is presented (framed) can influence how that information is processed and used in decision-making (Kapuściński & Richards, 2016).

Vehicle Selection

Your choice of communication vehicle depends primarily on which stage of the communication process you want to influence. Secondarily, the vehicles that you choose depend on the focus of your MarCom plan. Vehicles can often address more than one stage in the communication process but typically have a specific strength within a certain stage (Table 19.1).

Budget and Schedule

After you've identified the appropriate mix of communication vehicles, determine the costs of your plan, compare them to your budget, and adjust accordingly. The final step is to create a tactical implementation calendar that outlines when each vehicle is released and who is responsible for getting it done.

Establishing the Total Marketing Communications Budget

One of the hardest marketing decisions facing companies is how much to spend on promotion. John Wanamaker, the department store magnate, once said, "I know that half of my advertising is wasted, but I don't know which half. I spent $2 million for advertising, and I don't know if that was half enough or twice too much."

Table 19.1 Sample Case for MarCom

Case Study: The Renaissance of the Majestic Oasis Resort
Background:
The Majestic Oasis Resort, situated on the pristine beaches of Zanzibar, was once the epitome of luxury and the preferred destination for elite travellers. However, over the past few years, due to increased competition and a lack of effective marketing, the resort had seen a decline in occupancy rates and overall brand perception.
Objective:
To rejuvenate *The Majestic Oasis Resort's* image, regain its position as a premium luxury destination in the minds of international travellers, and increase occupancy rates by 20% over the next year.
Strategy:
A comprehensive MarCom strategy was formulated, centred around the unique selling proposition (USP) of the resort: "A Sanctuary of Timeless Luxury."
1. **Target Audience Profiles:**
Through market research, two primary audiences were identified:
- Affluent couples aged 30–50 years seeking romantic getaways.
- Families looking for luxury vacations with world-class amenities.
2. **Processes of Buying, Selling, and Communication:**
Research indicated that the target audience primarily used online platforms for holiday bookings, influenced by digital ads, influencers, and travel reviews.
Decisions Made:
1. **Objectives:**
- Reinforce the resort's image as a premier luxury destination.
- Increase direct bookings through the resort's website.
- Foster collaborations with influencers and travel bloggers.
2. **Messages:**
- Highlight the resort's unique experiences, such as private beach dinners, spa services, and cultural excursions.
- Emphasise the "Sanctuary of Timeless Luxury" in all communications.

(Continued)

Table 19.1 (Continued) Sample Case for MarCom

Case Study: The Renaissance of the Majestic Oasis Resort
3. **Vehicles:**
- Launch a redesigned website with immersive photo galleries, virtual tours, and guest testimonials.
- Collaborate with travel influencers to showcase real-life experiences at the resort.
- Initiate targeted digital ad campaigns on platforms frequented by the target demographic, like luxury travel blogs, Instagram, and YouTube.
4. **Budget and Schedule:**
- Allocate 40% of the budget to website redesign and maintenance.
- Dedicate 30% to digital advertising and retargeting campaigns.
- Reserve 20% for influencer collaborations.
- Set aside 10% for contingencies and any ad-hoc promotional activities.
- Roll out the redesigned website in Q1, initiate influencer collaborations in Q2, and ramp up digital advertising in Q3, leading up to the peak holiday season.
Results: *(projected)
Post-implementation of the MarCom plan:
- The Majestic Oasis Resort saw a 25% increase in direct bookings within the first six months.
- Occupancy rates increased by 22% year on year.
- Guest feedback was overwhelmingly positive, with many citing the compelling digital ads and influencer testimonials as their reason for booking.
- The resort regained its esteemed position in luxury travel forums and received accolades for its effective use of marketing communications in the hospitality sector.
Conclusion The renaissance of The Majestic Oasis Resort underscores the potency of a well-thought-out MarCom plan. By understanding its audience, leveraging modern communication vehicles, and staying true to its brand essence, the resort not only achieved but surpassed its objectives. It stands as a testament to the power of strategic marketing communications in the realm of hospitality and tourism.

Large budgets are not required for well-planned and well-executed communications. Pam Felix, owner of the California Tortilla restaurant in Bethesda, Maryland, states, "As an independent, I do not have much of an ad budget." Pam uses humour as the basis of her promotions. She claims, "The goofier we are the more money we make." Her main communication tool is a newsletter called Taco Talk. An example of one of her successful promotions was Jungle Noise Day. Everyone who came in and made a noise like Tarzan received free chips and salsa. There is a lack of parking in Bethesda, and as a result parking tickets are common. Felix gives anyone who comes in with a parking ticket a free taco to help relieve the pain of getting the ticket. She uses these wacky promotions to create a fun atmosphere and also as a way to give something back to her customers.

How do companies determine their promotion budget? Four common methods are used to set the total budget for advertising: (1) the affordable method, (2) the percentage of sales method, (3) the competitive parity method, and (4) the objective and task method.

Affordable Method

The affordable method is one where companies set their promotion budget based on what they believe the company can afford. Imagine a local bakery trying to decide its promotional budget. The owner might think, "I've had a good few months of sales, maybe I can set aside $2,000 for marketing this year." This approach is equivalent to a homeowner deciding to paint their house only if they have spare cash at the end of the year.

However, the affordable method tends to overlook the actual impact of promotion on boosting sales. Because it's arbitrary, it might lead to inconsistent yearly budgets, making it tough for long-term marketing planning. This method can sometimes lead to overspending but more often can cause companies to underinvest in advertising.

Percentage of Sales Method

This method involves companies setting their promotional budget as a percentage of their current or forecasted sales. Think of a fashion retailer setting aside 5% of their annual revenue for marketing. This could mean that if they make $1 million in sales, they'd allocate $50,000 for promotional efforts.

The benefits touted for this approach include its simplicity and the belief that promotional spending fluctuates with what a company can "afford." It's akin to you deciding to spend 5% of your monthly income on entertainment, regardless of any change in your salary. But this approach has pitfalls. It's like saying promotion leads to sales when it's the other way around. If sales decrease, this method can hinder the necessary increased promotional expenditure.

Competitive Parity Method

Here, companies set their promotion budgets in line with their competitors' expenditures. Imagine two rival coffee shops on the same street. If one decides to invest in a big advertising campaign, the other might feel the need to match or exceed that expenditure to stay competitive. If the industry norm for coffee shops is to spend 3 percent of sales on marketing, they might all try to adhere to that figure.

The two main arguments for this method are that it reflects industry wisdom and can deter promotional wars. However, these arguments are shaky at best. Businesses have unique needs and blindly mirroring competitors might not serve a company's specific goals or target audience.

Objective and Task Method

This method is perhaps the most systematic. Companies first define clear promotional objectives, identify the tasks needed to achieve those objectives, and then estimate the costs associated with those tasks. For instance, a bookstore wanting to increase its customer base by 20% might decide to host monthly author events. They'd then calculate the costs of these events—hiring the venue, paying the authors, promoting the event—and set that as their promotional budget.

The objective and task method requires companies to be explicit about their expectations between promotional spend and outcomes. It's like deciding to get fit, outlining a specific workout regimen, and then budgeting for gym membership, personal trainer fees, and diet plans. While it offers clarity, it can also be challenging due to the intricacies involved in identifying which tasks will deliver on the set objectives.

1. **Affordable Method**

Step by Step:

1. *Assess Finances*: The Chief Finance Officer (CFO) reviews the financial statements to determine the company's overall health.
2. *Set Aside Funds*: The CFO determines they have an excess of GHS 20,000 that can be allocated for marketing for the upcoming year.
3. *Allocate*: The marketing team gets the GHS 20,000 budget and now must decide how to spend it across various campaigns.

Example: *Based on current finances and after accounting for all operational costs, Green Cocoa feels they can "afford" to spend GHS 20,000 on marketing.*

2. **Percentage of Sales Method**

Step by Step:

1. *Review Sales*: Green Cocoa checks its total sales from the previous year and finds it made GHS 500,000.
2. *Determine Percentage*: They decide to allocate 4 percent of their sales to marketing.
3. *Calculate Budget*: Four percent of GHS 500,000 is GHS 20,000, setting their marketing budget.

**Example:* If Green Cocoa anticipates or targets sales of GHS 600,000 for the next year, at 4 percent, they will set aside GHS 24,000 for marketing.*

3. **Competitive Parity Method**

Step by Step:

1. *Research Competitors*: Green Cocoa investigates its major competitors in the cocoa processing industry and finds most spend around GHS 25,000 annually on marketing.
2. *Match or Adjust*: They decide to match the industry average, setting their budget at GHS 25,000.

3. *Allocate*: The budget is then divided among various campaigns to compete effectively.

Example: If one of Green Cocoa's direct competitors launched a major campaign last year and saw significant growth, Green Cocoa might even decide to allocate slightly more, say GHS 27,000, to try and match or surpass this success.

4. **Objective and Task Method**

Step by Step:

1. *Set Objectives*: Green Cocoa wants to expand its customer base by 15 percent and introduce a new cocoa product.
2. *List Tasks*: To achieve this, they decide to:
   a. Launch a digital ad campaign.
   b. Host a series of tasting events.
   c. Collaborate with influencers for promotion.

5. **Estimate Costs:** They find:
   a. Digital ads will cost GHS 8,000.
   b. Tasting events will cost GHS 7,000.
   c. Influencer partnerships will be GHS 10,000.

6. **Total the Costs**: Summing these up, they get a total of GHS 25,000 as their marketing budget.

**Example:* If midway, they decide to expand to another region in Ghana, they'd have to revisit this budget, re-estimate costs, and adjust accordingly.*

For most companies, it's not about rigidly sticking to one method. The real-world approach is often a blend. Green Cocoa, for example, might use the percentage of sales method to get a general figure, then refine this figure using the objective and task method to ensure their campaigns are adequately funded.

Summary

In this chapter, we examined the significance of structured marketing communications for product promotion. It began with an overview of crafting a comprehensive marketing communications plan and proceeded to address the integral components of budgeting and scheduling. The chapter systematically explored four principal methods for establishing a marketing communications budget: the Affordable Method, the Percentage of Sales Method, the Competitive Parity Method, and the Objective and Task Method. By the chapter's conclusion, readers have gained a holistic understanding of how to formulate a robust marketing communications plan and the nuances of budget allocation.

Activity

1. Outline the steps in drawing a marketing communications plan.
2. In profiling your target audience, what are the key demographic and psychographic variables you need to understand about your target audience?
3. Discuss the four common methods that managers in the hospitality industry could use to set the total budget for advertising.
4. Design a marketing campaign for Green Cocoa targeting European markets considering Green Cocoa's background and its commitment to sustainability, community empowerment, and the celebration of Ghana's cocoa heritage.

References

Itani, O. S., Badrinarayanan, V., & Rangarajan, D. (2023). The impact of business-to-business salespeople's social media use on value co-creation and cross/up-selling: The role of social capital. *European Journal of Marketing*, *57*(3), 683–717.

Kapuściński, G., & Richards, B. (2016). News framing effects on destination risk perception. *Tourism Management*, *57*, 234–244.

Keller, K. L. (2009). Building strong brands in a modern marketing communications environment. *Journal of Marketing Communications*, *15*(2–3), 139–155.

Kim, J. H. (2014). The antecedents of memorable tourism experiences: The development of a scale to measure the destination attributes associated with memorable experiences. *Tourism Management*, *44*, 34–45.

Nguyen, N. P., & Mogaji, E. (2022). Marketing communications strategies for public transport organisations. In *Public sector marketing communications volume I: Public relations and brand communication perspectives* (pp. 41–68). Cham: Springer International Publishing.

Chapter 20

Professional Sales

Chapter Outline

- Introduction
- Management of Professional Sales
- Nature of Hospitality Sales
- Sales Force Objectives
- Sales Force Tactics

Objectives

After reading this chapter, you should be able to:

- Explain the role and nature of personal selling in the hospitality industry.
- Describe the basics of managing the sales force, and tell how to set sales force strategy, how to pick a structure—territorial, product, customer, or complex—and how to ensure that sales force size is appropriate.
- Discuss supervising salespeople, including directing, motivation, and evaluating performance.
- Apply the principles of personal selling process and outline the steps in the selling process—qualifying, pre-approach and approach, presentation and demonstration, handling objections, closing, and follow-up.

DOI: 10.4324/9781032688497-20

Introduction

Everyone sells in some form or another. However, there are those who bear the direct responsibility of ensuring that businesses remain profitable, payrolls are managed, bills are settled, and investors see a reasonable return. These individuals are the backbone of professional sales. In this chapter, we delve deep into the nuances of professional selling, especially in the context of the hospitality sector, an industry that remains intrinsically tied to the ebb and flow of the economy.

Management of Professional Sales

The hospitality industry's success hinges on the mastery of selling. It might be tempting to think that a bustling roadside motel situated at a prime location or a buzzing restaurant with customers queuing up doesn't need to "sell." But that's a short-sighted perspective, one that no one in the hospitality industry should entertain for long.

Astute owners and managers recognise the continuous imperative to sell. Not just to customers, but to stakeholders of various shades: from county commissioners, tax officials, and planning boards to media professionals, bankers, and local tourism enthusiasts. Even the in-house team, be it those analysing credit card data, handling audio-visuals, administrative staff, or those ensuring the upkeep of the property, are integral to the selling process.

Stories abound of missed sales opportunities or strained guest relationships, often resulting from an oversight or a misstep by support staff who erroneously felt they were not part of the selling equation.

The term "sales representative" in the hospitality domain is multifaceted, spanning roles that might seem disparate but are united in their sales objective. These roles can be classified as:

- *Deliverer*: Those primarily tasked with product delivery, like suppliers for restaurants or hotels.
- *Order Taker*: Individuals, whether inside (like reservation desks) or outside (like suppliers liaising with chefs), primarily handling order placements.
- *Missionary*: Those not necessarily involved in order-taking but focused on relationship-building and education. Airline representatives liaising with travel agencies or those representing their brands at trade shows exemplify this role.

- *Technician*: Roles where technical know-how is paramount. This includes salespersons offering consultancy based on specialised knowledge, perhaps in yield management to entities like hotels or airlines.
- *Demand Creator*: Those involved in the proactive selling of tangible products or services, which is characteristic of a significant portion of the hospitality industry.

These roles showcase the spectrum of selling, from reactive order servicing to proactive, creative customer engagement. We will dwell in this chapter on issues around understanding the nature of hospitality sales, setting the objectives for the sales force, structuring and sizing the sales force, organising the sales department for optimal functionality, mastering the nuances of recruiting and training a top-tier sales force, and effective management strategies for overseeing the sales force.

Nature of Hospitality Sales

Sales personnel serve as a hospitality institution's link to customers. The sales representative is the company to many customers and in turn brings back the much-needed customer intelligence. Personal selling can be one of the most expensive contact and communication tools used by hospitality and tourism institutions.

Cost estimates for making a personal sales call vary depending on the industry and the company, but one conclusion remains constant. However, as measured, the cost is high! A *Business Week* report noted that "There are 8 million people in the U.S. workforce who are directly involved in sales, and it now costs $250 and up to send any one of them on a call." A no-hospitality company, E. I. DuPont de Nemours and Company, Inc., estimates a cost-per-field sales call of $5008 and above, and an even higher estimate of $700 per visit is given by a researcher who included the salesperson's salary, cost of travel, technical support people, and cost of presentations. Added to this the fact that sales orders are seldom written on the first call and often require five or more calls, particularly for larger orders. The cost of obtaining a new client thus becomes enormously high.

Cost Implications of Personal Sales Calls in the African Context

Costs related to personal sales calls exhibit variations across industries and countries, yet there is a common theme: they represent a significant investment for businesses. An insightful report by *African Business Magazine* states, "Across several African nations, a considerable portion of the workforce is directly involved in sales, with the average cost of sending a representative on a call ranging from $100 to $200, depending on the region and industry."

For instance, Safaricom, a leading telecommunication company in Kenya, reported that its cost-per-field sales call for business solutions ranged between $150 and $200. This estimate factors in the salesperson's salary, travel allowances, and training costs. In Nigeria, Dangote Industries, a conglomerate with interests in commodities, suggested costs of around $180 for their sales reps visiting potential B2B clients, which includes their transportation, accommodation in certain cases, and support materials.

However, these costs can escalate further when we consider the complexities of sales in certain sectors. In South Africa, for example, companies in the mining equipment sector have mentioned that multiple visits are often required to finalise deals, especially with large corporations or international buyers. Securing a significant contract in such industries can necessitate up to seven meetings, driving the overall client acquisition cost exponentially higher in some African countries (McKinsey & Company, 2022).

In the African context, sales representatives play crucial roles given the diverse cultural, economic, and geographic landscapes. They navigate various challenges, from understanding localised needs to building trust in communities. Their tasks often extend beyond mere sales pitches to include:

1. *Prospecting*: Sales representatives find and cultivate new customers.
2. *Targeting*: Sales representatives decide how to allocate their scarce time: among prospects and customers.
3. *Communicating*: Sales representatives communicate information about the company's products and services.
4. *Selling*: Sales representatives know the art of salesmanship: approaching, presenting, answering objections, and closing sales.
5. *Servicing*: Sales representatives provide various services to the customers—consulting on their problems, rendering technical assistance, arranging financing, and expediting delivery.
6. *Information Gathering*: Sales representatives conduct market research and intelligence work and fill-in call reports.
7. *Allocating*: Sales representatives decide which customers to allocate scarce products to during product shortages.

The Evolving Role of Sales Representatives in the Global and African Context

The responsibilities and challenges facing sales representatives are directly influenced by economic conditions and market dynamics. During product shortages, such as a temporary shortage of hotel rooms during a major convention, sales representatives find themselves with nothing to sell (Storbacka et al., 2009; Christ and Anderson, 2011; Cuevas, 2018). Some companies jump to the conclusion that fewer sales representatives are then needed. But this thinking overlooks the salesperson's other roles—allocating the product, counselling unhappy customers, and selling the company's other products that are not in short supply. It also ignores the long-run nature of hospitality sales.

Let's take an instance where when there's an unexpected surge in tourists during an African football championship or a major cultural festival like Ghana's PANAFEST, hotel rooms can become scarce. This can lead sales representatives to momentarily feel they have little to sell. Some businesses might prematurely conclude that in such times, fewer sales personnel are required. However, this line of thought neglects the multifaceted roles that salespeople play, which include product allocation, customer relationship management, and promoting alternative services or products still available. Furthermore, this short-sighted view undermines the long-term aspects of the hospitality sales process.

Many conventions and conferences are planned years, and hospitality salespeople must often work with meeting and convention planners two to four years in advance of the actual event. Resorts in the United States have concentrated much of their selling efforts on meetings and conferences, which now represent 35 percent of their customers. This was not achieved by viewing professional sales as a short-run tactic. A senior analyst with Tourism Canada has demonstrated that Canadian resort salespeople are effective in reaching foreign markets. Guests in Canadian resorts are 60 percent Canadian and 40 percent foreign. By comparison, US resorts have a mix of 91 percent American and 9 percent foreign.

Similarly, South Africa's Sun City Resort and other prominent hotels often report that a significant portion of their bookings come from events and conventions, which can account for as much as 40% of their clientele. This kind of business planning isn't achieved with a short-term sales mindset. Research from the South African Tourism Board indicates that while domestic guests make up a large portion of resort clientele, there's an increasing number of visitors from other African nations and globally. This shift emphasises the need for a long-term, strategic approach to sales, especially when targeting foreign markets.

Again, Nigeria's Eko Hotels & Suites in Lagos has also seen a rise in international events, driving global clientele. The emphasis is clear: penetrating international markets requires foresight, dedication, and strategic long-term planning.

As companies move towards a stronger market orientation, their sales forces need to become more market-focused and customer-oriented (Mulki et al., 2008). The traditional view is that salespeople should worry about volume and "sell, sell, sell," and that the marketing department should worry about marketing strategy and profitability. The newer view is that salespeople should know how to produce customer satisfaction and company profit. They should know how to analyse sales data, measure market potential, gather market intelligence, develop marketing strategies and plans, and become proficient at the use of sales tactics.

Sales representatives need analytical skills: This becomes especially critical at the higher levels of sales management. Marketers believe that a sales force will be more effective in the long run if members understand marketing and selling. The newer concept is basic to the successful use of yield management in the hospitality industry.

This has become very clear as database marketing has gained importance within the hospitality industry. Group sales have been particularly affected. In the contemporary business landscape, both in Africa and globally, companies are gravitating towards a more market-centric approach. Sales teams, in turn, are expected to be not only target-driven but also market-informed and customer-focused. The outdated perception that sales teams should solely be focused on pushing volumes is fading. Today's sales professionals, whether in Kenya's bustling Nairobi or New York's metropolis, are expected to understand how to deliver customer satisfaction while simultaneously driving company profit. They're now equipped with skills to interpret sales data, assess market potential, gather valuable market insights, develop strategic marketing plans, and execute sales tactics effectively.

Sales Force Objectives in the Hospitality Industry

Every successful venture in the hospitality industry understands the need to set clear, actionable objectives for its sales force. These objectives serve as guiding principles for several reasons:

1. *Alignment with Corporate Goals*: Objectives ensure the alignment of sales strategies with overarching corporate goals. These goals can span various areas–from revenue targets and market share expansion to brand enhancement and reputation management.
2. *Guidance for Sales Team*: Clearly defined objectives offer the sales force a roadmap to plan and implement their sales initiatives. They also ensure optimum utilisation of resources, be it time or tools like personal computers and software.

While sales objectives are tailored annually for individual companies, they usually resonate with larger corporate goals, which are then delineated into quarterly and monthly targets (Westbrook and Peterson, 2022). For the sales force, these are further broken down into weekly or even daily milestones.

The onus of setting and communicating these objectives typically lies with the sales manager, often crafted in collaboration with the salesperson. Given that seasoned salespeople possess first-hand market insights, their feedback is invaluable in shaping realistic, achievable objectives.

There might be instances when annual objectives need modification before the year concludes. Such changes can arise from unforeseen events, whether it's geopolitical issues like wars, natural calamities like floods, significant economic shifts like currency fluctuations, or internal changes such as hotel acquisitions.

Despite the tailored nature of sales objectives, there are some general objectives frequently adopted by the hospitality sector:

Sales Volume: Metrics like occupancy rates for hotels, passenger miles for airlines, or covers for restaurants, all hint at a singular objective: maximise customer inflow. However, a sole focus on volume can sometimes lead to pitfalls like excessive price cuts, catering to less-profitable customer segments, and employee dissatisfaction due to increased workloads without corresponding benefits.

Segment-specific Sales Volume: Premium establishments, like luxury resorts or high-end cruises, often adopt a philosophy centred on volume objectives but with a caveat. They focus on specific, affluent customer segments, believing that maintaining quality and exclusivity ensures sustained profitability. For example, a luxury lodge in the Maasai Mara, such as "Mara Engai," might focus on attracting high-end tourists by offering exclusive safari packages. They believe in the principle of exclusivity ensuring sustained profitability. While such a strategy may be effective for niche establishments, a broader chain like "Protea Hotels by Marriott" would have varied strategies across their different hotel categories.

While this strategy may be effective for niche establishments, it's not a one-size-fits-all solution for the broader hospitality industry. Nonetheless, the idea of setting segment-specific sales objectives is pivotal for a targeted and efficient sales approach, and these strategies should be continually revised based on sales performance data.

Sales Force Tactics

We briefly examine eight major aspects of personal selling: prospecting and qualifying, pre-approach, approach, presentation and demonstration, negotiation, overcoming objections, closing, and follow-up maintenance.

1. ***Prospecting and Qualifying***

The first step in the selling process is to identify prospects. Although the company will try to supply leads, sales representatives need skills in developing their own. Leads can be developed in the following ways:

1. Through call-ins.
2. Having a booth at appropriate travel or trade shows.
3. Participating in international travel missions.
4. Asking current customers for the names of prospects.
5. Cultivating other referral sources, such as suppliers, dealers, non-competing sales representatives, bankers, and trade association executives.
6. Through leads generated by the chain.
7. Joining organisations to which prospects belong.
8. Engaging in speaking and writing activities that will draw attention.
9. Examining data sources (newspapers, directories) in search of names.
10. Using the telephone and mail to find leads.
11. Dropping in unannounced on various offices (cold canvassing).
12. Online and social media campaigns.
13. Conducting a sales blitz.

Front desk staff could delicately prospect guests who patronise H&T institutions. Prospecting internally and externally should be done daily. Once prospects have been identified, they need to be qualified.

Sales representatives need skills in screening out poor leads. Prospects can be qualified by examining their financial ability, the volume of business, special requirements, location, and the likelihood of continuous business. The salesperson might phone or write to prospects before deciding whether to visit them. Leads can be categorised as hot leads, warm leads, and cool leads.

2. ***Pre-approach***

The salesperson needs to learn as much as possible about the prospective company (what it needs, who is involved in the purchase decision) and its buyers (their characteristics and buying styles). The salesperson should set call objectives, which might be to qualify the prospect, gather information, or make an immediate sale. Another task is to decide on the best approach, which might be a personal visit, a phone call, or a letter. The best timing should be thought out because many prospects are busy at certain times. Finally, the salesperson should plan an overall sales strategy for the account.

3. ***Approach***

The salesperson should know how to greet the buyer to get the relationship off to a good start. This involves the salesperson's appearance, the opening lines, and the follow-up remarks. The opening line should be positive, for example, "Mr. Smith, I am Alice Jones from the ABC Hotel Company, my company and appreciate your willingness to see me. I will do my best to make this visit profitable and worthwhile for you and your company." This might be followed by key questions and active listening to understand the buyer and his or her needs.

4. ***Presentation and Demonstration***

The salesperson now tells the product's "story" to the buyer, following the AIDA formula of gaming attention, holding interest, arousing desire, and obtaining action. The salesperson emphasises customer benefits throughout, bringing in product features as evidence of these benefits. A benefit is any advantage, such as lower cost, less work, or more profit for the buyer. A feature is a product characteristic, such as weight or size. A common selling mistake is to dwell on product features (a product orientation) instead of customer benefits (a market orientation).

Sales presentations can be improved with demonstration aids such as booklets, flipcharts, slides, and videos. During the demonstration, the salesperson can draw on five influence strategies:

1. *Legitimacy*: The salesperson emphasises the reputation and experience of his or her company.
2. *Expertise*: The salesperson shows a deep knowledge of the buyer's situation and the company's products, doing this without being overly "smart."
3. *Referent Power*: The salesperson builds on any shared characteristics, interests, and acquaintances.
4. *Ingratiation*: The salesperson provides personal favours (a free lunch, promotional gratuities) to strengthen affiliation and reciprocity feelings.
5. *Impression*: The salesperson manages to convey favourable personal impressions.

5. ***Negotiation***

Much of business-to-business selling involves negotiating skills. The two parties need to reach an agreement on the price and other terms of sort. Salespersons need to win the order without making deep concessions that will hurt profitability.

Although price is the most frequently negotiated issue, other issues include quality of goods and services offered, purchase volume, responsibility for financing, risk-raking, and promotion. The number of negotiation issues is virtually unlimited.

Unfortunately, far too many hotel salespeople rely almost exclusively on price as their negotiating tool. Even worse, they often begin negotiating from an already discounted price rather than from rack rates. Negotiations should always begin with rack rates, and price concessions should be given only when essential. Numerous bargaining tools exist, such as upgrades, complimentary tickets for the ski lift or golf courses, first-class coffee breaks instead of coffee and soft drinks, airport pickup, and the use of hotel services such as the fitness centre. A hotel sales force might package these amenities into bundles of services and give them names such as the President's Package, the Connoisseurs Package, and the Executive Package.

Sales force members should be taught to negotiate using services or bundled services as the primary negotiating tool rather than price.

6. ***Overcoming Objections***

Customers almost always pose objections during the presentation or when asked for the order. Their resistance can be psychological or logical. Psychological resistance includes resistance to interference, preference for established hotels or airlines, apathy, reluctance to give up something, unpleasant associations about the other person, predetermined ideas, dislike of making decisions, and neurotic attitude towards money. Logical resistance might consist of objections to the price or certain product or company characteristics. To handle these objections, the salesperson maintains a positive approach, asks the buyer to clarify the objection, denies the validity of the objection, or turns the objection into a reason for buying. The salesperson needs training in the broader skills of negotiation, of which handling objections is a part.

7. ***Closing***

Now the salesperson attempts to close the sale. Some salespeople do not get to this stage or do not get it well. They lack confidence or feel uncomfortable about asking for the order or do not recognise the right psychological moment to close the sale. Salespersons need to know how to recognise closing signals from the buyer, including physical actions, statements or comments, and questions. Salespersons can use one of several closing techniques. They can ask for the order, recapitulate the points of agreement, offer to help the secretary write up the order, ask whether the buyers want A or B, get the buyer to make minor choices such as on colour or size, or indicate what the buyer will lose if the order is not placed now. The salesperson might offer the buyer specific inducements to close, such as a special price.

A basic problem mentioned over and over by hotel sales managers is that some members of the sales force do not ask for the order. They may follow all the other steps to perfection but for some reason seem incapable of asking for order.

8. ***Follow-up/Maintenance***

This last step is necessary if the salesperson wants to ensure customer satisfaction and repeat business. Immediately after closing, the salesperson should complete any necessary details on delivery time, purchase term, and other matters. Follow-up or foul-up is a slogan of most successful salespeople. The salesperson should develop an account maintenance plan to make sure that the customer is not forgotten or lost.

The sales culture has long been considered unique to other organisational units. Differences in issues such as how professionals are socialised, work performance expectations (Mulki, Lassk, Jaramillo, 2008), job behaviours (Jaramillo, Mulki, Boles, 2011), and pay structures (e.g., commission, bonus, combination plans) all create a unique environment in which work conditions, structure, and expectations vary from most other organisational units.

Summary

In this chapter, we explained the role and nature of personal selling and the role of the sales force. We also described the basics of managing the sales force, and tell how to set a sales force strategy, how to pick a structure—territorial, product, customer, or complex—and how to ensure that the sales force size is appropriate.

Activity

1. Explain the nature and role of personal selling, and its role in tourism and hospitality marketing.
2. Discuss the explain selling process and state what the salespeople need to do under each stage.
3. Discuss the basics of managing the sales force and how to set sales force strategy.
4. Explain how managers in the hospitality industry can effectively structure the sales force, design territory effectively, and ensure that the sales force size is appropriate.

References

Bodie, Z., Kane, A., & Marcus, A. (2013). *Ebook: Essentials of investments: Global edition*. McGraw Hill.

Cuevas, J. M. (2018). The transformation of professional selling: Implications for leading the modern sales organization. *Industrial Marketing Management, 69*, 198–208.

Christ, P., & Anderson, R. (2011). The impact of technology on evolving roles of salespeople. *Journal of Historical Research in Marketing, 3*(2), 173–193.

Jaramillo, F., Mulki, J. P., & Boles, J. S. (2011). Workplace stressors, job attitude, and job behaviors: Is interpersonal conflict the missing link? *Journal of Personal Selling & Sales Management, 31*, 339–356.

McKinsey & Company. (2022). FinTech in Africa: The end of the beginning. https://www.mckinsey.com/industries/financial-services/our-insights/fintech-in-africa-the-end-of-the-beginning

Mulki, J. P., Lassk, F. G., & Jaramillo, F. (2008). The effect of self-efficacy on salesperson work overload and pay satisfaction. *Journal of Personal Selling & Sales Management, 28*, 283–295.

Storbacka, K., Ryals, L., Davies, I. A., & Nenonen, S. (2009). The changing role of sales: Viewing sales as a strategic, cross-functional process. *European Journal of marketing, 43*(7/8), 890–906.

Westbrook, K. W., & Peterson, R. M. (2022). Servant leadership effects on salesperson self-efficacy, performance, job satisfaction, and turnover intentions. *Journal of Business-to-Business Marketing, 29*(2), 153–175.

Chapter 21

Internal Marketing

Chapter Outline

- Introduction
- Internal Marketing
- Linked Employee and Customer Satisfactions
- The Internal Marketing Process
- Service Culture Establishment

Objectives

Upon completion of this chapter, you will possess the knowledge and skills necessary to perform the tasks discussed:

- Provide an explanation of the significance of internal marketing within a marketing programme.
- Define service culture and elaborate on its significance in having a company that prioritises customer service.
- Outline and explain the step-by-step process required to implement an internal marketing programme.

Introduction

Internal marketing is a strategic approach that focuses on treating employees as internal customers and promoting the organisation's goals, values, and

DOI: 10.4324/9781032688497-21

products or services to them. It involves applying marketing principles and techniques to create a positive and engaging workplace culture, fostering employee satisfaction and loyalty, and aligning employee behaviour with the organisation's overall objectives. Internal marketing goes beyond traditional human resources practices and seeks to empower employees by considering them as vital stakeholders in the organisation's success. Internal marketing operates on the principle that happy and involved employees are more likely to deliver outstanding customer experiences and represent the brand as ambassadors. Internal marketing strategies encompass various elements, including effective communication, employee training and development, recognition and rewards, transparent leadership, and a supportive work environment. Organisations can gain a competitive edge, reinforce their brand, and achieve long-term success in today's ever-changing business environment by investing in internal marketing initiatives.

Internal Marketing

When it comes to receiving unsatisfactory service at a hotel or restaurant, everyone has a unique personal experience that they can share. For example, John Tscholl, a customer service expert, expressed his disappointment with the Marriott Courtyard Inn. He was not notified about his father-in-law's heart attack, which was an emergency, and was also not woken up the next morning as scheduled. This is a clear failure on the part of the hotel to provide basic services and support to their guests. The dissatisfied John shared his negative experience at a Marriott Courtyard Inn with thousands of people, including potential customers of the Marriott brand. What is the lesson here? The lesson is that although Marriott invested so much resources and effort in creating the Courtyard concept and providing good physical facilities, it is not enough to ensure a great customer experience. The behaviour and performance of the employees also play a crucial role in whether guests will choose to return.

Having excellent customer service is vital for any organisation's competitive advantage, and internal marketing (IM) plays a crucial role in achieving this, regardless of the industry or business. Internal marketing involves treating employees like customers and communicating effectively with internal markets. Mishra (2010) emphasises the importance of this approach.

According to Mishra (2010), the success of a business depends on its employees, who should be educated, motivated, and retained in the organisation to ensure global competitiveness. In service-oriented businesses, frontline employees are crucial to the organisation's success as they directly interact with external customers.

Not all stories are negative, especially when it comes to travel experiences. In fact, many travellers have positive stories to share about exceptional service from hotel employees. The following examples demonstrate how hotels can keep customers coming back by addressing problems and acknowledging employees who go above and beyond for the hotel. Barry Urquhart recently shared an account of receiving exceptional service from a hotel during his stay. He promptly reported a guest who had accidentally locked himself out on his balcony, and the hotel went beyond the call of duty by rewarding him for his assistance. This unparalleled experience left him with a remarkable impression of the hotel.

Karl Albrecht, one of the authors of Service America, needed to hold a business meeting in his hotel room during his stay in Sydney. Despite the hotel being fully booked and needing the room for arriving guests, he requested a late checkout. The manager was quick to accommodate him and provided a conference room free of charge. Although Albrecht was prepared to pay, the hotel declined, recognising the value of his future business and positive word-of-mouth, which outweighed any lost revenue. This explains the importance of employees in customer satisfaction. For example, when faced with a challenge, the hotel manager went above and beyond to accommodate their guest's needs—recognising the potential benefits of future business and positive word-of-mouth. The act of kindness and excellent customer service showcases the importance of valuing and prioritising relationships over short-term profits.

In order to succeed in the highly competitive hospitality and travel industry, it is important to encourage all employees to actively participate in marketing efforts. That is, rather than simply relying on the marketing or sales department, every member of the organisation should be fully integrated into the philosophy of marketing. This is particularly important in service industries, where line employees play a critical role in shaping customers' experiences and perceptions of the brand. To ensure that all employees are engaged in the marketing process, it is essential to implement internal marketing strategies that target the company's internal customers—its employees. By fostering a culture of marketing excellence and empowering all employees to contribute to marketing efforts, it becomes much easier for companies to achieve competitive advantage in the highly competitive hospitality and travel industries.

Managers should realise the significance of negative service encounters as they tend to receive more attention than positive ones. If a customer is not treated well, they are likely to share their negative experience with others. According to a study conducted by the Technical Assistance Research Program, people usually share their good experiences with an average of five people. In comparison, bad experiences are shared with an average of ten people. This makes it quite challenging to spread positive word of mouth, especially with social media has influence on our lives today. If a customer has a negative experience, they are likely to share it on various social media platforms, which can offset the impact of positive experiences. Therefore, the main objective should always be to fulfil or surpass every guest's expectations to ensure that they have a positive experience, which will encourage them to share their experience in a positive light.

The level of satisfaction that a guest experiences during their stay at a hotel is heavily influenced by their interactions with various hotel employees, such as the receptionist, those providing catering services, and room services. The appearance, attitude, and willingness of these employees to cater to the guest's requests play a significant role in forming an impression of the hotel and contributing to the guest's overall experience. Due to the nature of the hospitality industry, employees are responsible for delivering the products and becoming an integral aspect of product delivery. It is often challenging to differentiate tangible aspects of competing companies' products, such as steak dinners and hotel rooms in the same price range. However, product differentiation can often be derived from the service employees who deliver the service. Within the hospitality industry, non-marketing staff are often responsible for executing marketing activities instead of the designated marketing department. As a result, the hotel's marketing programme is heavily dependent on the ability of these employees to draw in guests to the establishment.

The primary objective of the hotel's staff is to foster customer loyalty and encourage guests to return for future stays. This is because research has shown that even a modest 5 percent increase in customer retention can translate into a substantial 25 percent uptick in the hotel's overall profits. Consequently, the satisfaction of employees, particularly those who interact directly with customers, is of paramount importance. These employees are pivotal in shaping the guest experience and their overall satisfaction levels (Greene et al., 1994).

The Link between Employee Satisfaction and Customer Satisfaction

Investing in a positive internal programme can have a significant impact on employee satisfaction, which in turn translates to happier customers. This was a core belief of J.W. Marriott, the founder of Marriott Hotels, who emphasised treating employees with the same level of care as customers. The benefits of internal marketing extend to both employee and customer satisfaction, leading to lower employee turnover rates and a positive impact on the bottom line. A study conducted by Forte Hotels in London found that well-trained and content employees tend to be loyal with their employer for longer periods. It's important to remember that interacting with dissatisfied customers can negatively affect employee morale and ultimately lead to higher turnover rates, as noted by Berry and Parasuraman (2004). Numerous studies have found a strong link between the happiness of customers and employees. This connection works both ways, meaning that the higher the satisfaction of employees, the higher the possibility that customers satisfaction will also be achieved.

Experts like Reichheld and Owen have proved that employee retention is closely linked to customer retention and satisfaction. Akroush et al. (2013) identified six dimensions of IM, with staff recruitment, training, and internal communication positively impacting staff motivation. Papadas et al. (2019) also suggest that internal green marketing can play a role in achieving a competitive advantage in a highly competitive market. Overall, it is clear that retaining employees is crucial for maintaining customer satisfaction and achieving business success.

The Internal Marketing Process

For businesses to excel at providing their customers with exceptional service, they must prioritise encouraging their employees to perform at their best. To achieve this goal, implementing internal marketing strategies is key. By adopting formalised procedures for marketing to employees, businesses can ensure that every employee understands their objectives and campaigns through a customer-focused lens. Internal marketing adopts a marketing perspective, motivating employees to deliver outstanding service and performance that prioritises the customer's needs. According to Christian Gronroos, a coordinated, marketing-like approach that involves various activities can achieve this goal. Ultimately, implementing internal marketing strategies helps businesses manage their employees effectively, ensuring they deliver the best possible service to their customers.

Internal marketing is a vital strategy that aims to boost a company's growth by directing marketing efforts towards its employees. This process involves:

- Fostering a service-oriented culture within the organisation.
- Adopting a marketing-driven approach to human resource management.
- Providing marketing information to employees.
- Implementing a reward and recognition programme that acknowledges good work and encourages employees to strive for excellence.

By focusing on internal marketing, companies can increase employee engagement, motivation, and productivity, leading to improved business outcomes and increased profitability.

Service Culture Establishment

For a service marketing programme to be successful, it must stem from a service culture within the organisation. Recently, an Australian national newspaper reported that despite investing $2 million into customer service programmes, four firms failed to produce significant results. The reason for this failure was the lack of a service-oriented culture within these companies. They implemented customer service programmes solely to generate satisfied customers and increase profits. However, they soon realised that effective customer service requires a deeper commitment from the management, not just line employees. Therefore, an internal marketing programme can only thrive if management is fully committed to a service culture.

Internal marketing programmes often encounter a significant challenge in the form of middle management. Many managers prioritise cost-cutting and profit maximisation, with their incentive structures tied to meeting specific cost targets. For instance, consider a scenario in a hotel where front-desk clerks have undergone training and are enthusiastic about delivering excellent service to guests. They may spend more time with customers or offer complimentary access to the health club to address any issues faced by dissatisfied guests. However, the front-office manager, who hasn't received the same training, might perceive these additional efforts as unproductive and wasteful, conflicting with their focus on cost reduction and meeting financial targets. This misalignment between the front-line staff's dedication to service and middle management's cost-driven approach can hinder the successful implementation of internal marketing initiatives.

Organisations must recognise that exceptional customer service from employees requires the right support and encouragement from management. While one-day training sessions can be beneficial, their impact might not be long-lasting without ongoing support. Merely instructing receptionists to be helpful and friendly is insufficient if they are understaffed and overwhelmed. For instance, the typical greeting used by receptionists, such as "Good morning, Plaza Hotel, Elizabeth speaking, how may I help you?" can come across as insincere and unhelpful when rushed and followed by a quick request to hold. From the guest's perspective, waiting 14rings for the phone to be answered and then receiving a cold and hurried greeting can be disheartening.

To foster a strong service culture, the management must implement comprehensive policies, procedures, reward systems, and actions that prioritise and reinforce the importance of exceptional customer service. Adequate staffing, ongoing training and coaching, and the establishment of a supportive work environment are essential elements to ensure employees can consistently deliver outstanding customer experiences. By providing the necessary resources and creating an environment where employees feel empowered and valued, organisations can cultivate a culture of exceptional service that benefits both employees and customers.

A company's organisational culture is an indispensable element that contributes significantly to its success. It encompasses the shared values and beliefs that shape the company's identity and provide a sense of direction for its members. In essence, it establishes the rules for behaviour, ensuring everyone is on the same page regarding the expectations and standards for conduct.

In a well-managed organisation, every member embraces the culture and is committed to upholding its principles. This is important because a strong culture serves two critical purposes. Firstly, it guides behaviour, especially for service organisations with varying customers and experiences. Employees must have the discretion necessary to create and deliver unique experiences that meet customers' diverse needs and expectations. This requires a shared understanding of what is expected and unacceptable; culture can provide this guidance.

Secondly, a robust culture instils a sense of purpose in employees and fosters pride in the company. When employees understand the company's goals and their role in achieving them, they are more likely to be motivated and committed to their work. This creates a positive work environment and increases productivity and job satisfaction. In conclusion, a well-defined organisational culture is critical for any company that aims to succeed and create a positive work environment for its employees (Belias et al., 2020).

An organisation's culture is the binding force that unites its employees. A strong culture fosters unity and cooperation among employees. However, having a strong culture doesn't necessarily imply having a service culture. On the other hand, a service culture motivates employees to adopt customer-oriented behaviour, which is the initial step towards establishing a customer-centric organisation.

To build a customer-focused organisation, everyone involved needs to be committed. This requires changes in recruitment, training, reward systems, customer complaint handling, and empowering employees. Managers need to communicate with customers and staff who interact with them. Commitment from management is vital as a service culture is not created by a memo from the CEO. It's developed over time through management's actions. For instance, a hotel manager who greets guests, and checks on their well-being during morning checkout and afternoon check-in, shows that they care about guests. A study by Bailey et al. (2016) involving 234 employees in Saudi Arabian financial institutions found a positive correlation between internal marketing and organisational commitment.

Summary

This chapter has highlighted the significance of internal marketing in a marketing programme. Additionally, it has explored the concept of a service culture and the benefits of having a company that prioritises customer service. Finally, it has outlined a four-step process for implementing an effective internal marketing programme.

Activity

1. Define internal marketing and provide any three reasons why the concept is critical to success in the hospitality industry.

2. Define organisational culture and evaluate two reasons why a strong culture within an organisation is important.
3. Critically evaluate four means through which you can achieve internal marketing as an executive in a hospitality industry.
4. Describe the four-step process involved in implementing an internal marketing programme.
5. How important is service culture to the tourism industry in Africa?

References

Akroush, M. N., Abu-ElSamen, A. A., Samawi, G. A., & Odetallah, A. L. (2013). Internal marketing and service quality in restaurants. *Marketing Intelligence & Planning, 31*(4), 304–336.

Bailey, A. A., Albassami, F., & Al-Meshal, S. (2016). The roles of employee job satisfaction and organizational commitment in the internal marketing-employee bank identification relationship. *International Journal of Bank Marketing, 34*(6), 821–840. https://doi.org/10.1108/IJBM-06-2015-0097

Belias, D., Vasiliadis, L., & Velissariou, E. (2020). Internal marketing in tourism: The case of human resource empowerment on Greek hotels. In *Cultural and tourism innovation in the digital era: Sixth international IACuDiT conference, Athens 2019* (pp. 559–573). Springer International Publishing.

Berry, L. L., & Parasuraman, A. (2004). *Marketing services: Competing through quality.* Simon and Schuster.

Greene, W. E., Walls, G. D., & Schrest, L. J. (1994). Internal marketing: The key to external marketing success. *Journal of Services Marketing, 8*(4), 5–13.

Mishra, S. (2010). Internal marketing-A tool to harness employees' power in service organizations in India. *International Journal of Business and Management, 5*(1), 185.

Papadas, K. K., Avlonitis, G. J., Carrigan, M., & Piha, L. (2019). The interplay of strategic and internal green marketing orientation on competitive advantage. *Journal of Business Research, 104*, 632–643.

Chapter 22

Building Customer Loyalty through Quality

Chapter Outline

- Introduction
- Defining Customer Value and Satisfaction
- Customer-Delivered Value
- Service Quality
- Customer Satisfaction
- Relationship Marketing
- Relationship Marketing vs. Traditional Marketing
- Retaining Customers
- Cost of Lost Customers
- The Link Between Marketing and Quality
- What Is Quality?
- Benefits of Service Quality
- Customer Retention
- Avoidance of Price Competition
- Retention of Good Employees
- Cost Reduction

DOI: 10.4324/9781032688497-22

Objectives

After reading this chapter, you should be able to:

- Define customer value and customer satisfaction.
- Distinguish between customer satisfaction and customer loyalty.
- Discuss how to attract new customers and retain current customers through relationship marketing.
- Explain the tactics for resolving customer complaints and the importance of resolving such complaints.
- Define quality and discuss its importance to the tourism and hospitality business.
- Implement capacity and demand management tactics.

Introduction

The hospitality and tourism industry is facing tough competition, which is expected to intensify. To succeed, companies must prioritise meeting and satisfying customer needs. This chapter will delve deeper into how companies can win customers and outperform their competitors. The key to success lies in being customer-centred, delivering superior value to target customers, building customer relationships rather than just products, and being skilled in market engineering. Adapting to this new philosophy is essential for companies to survive and thrive in the current market. We understand it can be challenging, but the chapter provides some insight that will help design effective strategies and programmes to create value for customers.

Understandably, some companies may think that it's the sole responsibility of their marketing or sales department to gain new customers. However, truly successful companies know that this is not the case. While marketing plays an important role, it can only be described as a lead partner when it comes to attracting and retaining customers. Ultimately, all departments and employees must unite to create a customer value delivery system that stands out. After all, even the best marketing in the world cannot sell a poorly made product that doesn't meet the needs of consumers. According to Nobar et al. (2018), customer loyalty is a key factor in determining the success of hotels and tourism companies. Focusing on your strengths and resources can improve your supply and create more opportunities to build a strong reputation and increase your visibility. Remember, working as a team is key to achieving success!

Defining Customer Value and Satisfaction

It can be tough for companies to attract customers, as so many options are available to people these days. Peter Drucker recognised this over 35 years ago when he highlighted the importance of customers to businesses. With so many products, brands, prices, and suppliers out there, it's no wonder that customers sometimes struggle to decide what to buy. This is a key issue for companies to address, and we understand the challenges they face in doing so.

As customers, we all want the best value for our money, but it's not always easy to find. We must consider our search costs, knowledge, mobility, and income when deciding. And when we do make a choice, we have certain expectations of what we'll get in return. It's only natural to compare what we expected to what we actually received. This is important because it affects our satisfaction and our likelihood of returning. As managers, it's crucial to understand that these expectations can vary depending on the type of business. For example, when travelling to Hong Kong, we might have different expectations for the service at airport restaurants. Full-service restaurants might have the highest expectations, but unfortunately, all types of restaurants may fall short. That's why it's so important for managers to strive to meet or exceed the service expectations of their specific enterprise. Understanding customer value and satisfaction is key to running a successful business.

Customer-Delivered Value

As a customer, the value you get from a product depends on how well it meets your needs. The profit you gain from a marketing offer is the difference between its total cost and value, also known as customer-delivered value. This value includes all your benefits from a marketing offer, such as customer service, product quality, and overall image. On the other hand, customer cost includes all the monetary, time, energy, and psychological costs of a marketing offer.

For example, if you're a business traveller, you may prefer a non-stop flight instead of a direct flight with a stopover, as it saves you time. You may also avoid airports that require a lot of walking and choose specific connecting points. Additionally, you may prefer airlines with good on-time and customer service records, as this reduces your mental stress. These attributes add value to your overall experience. Let's say an airline offers a non-stop flight for $25 more than a competitor's flight with a stopover in Dallas, which takes two hours longer. The hassle of walking through the airport and the extra two hours increase your total customer cost, even though the non-stop flight costs more. However, in this case, you would still prefer the non-stop flight because it offers you a higher customer-delivered value.

A comprehensive survey was conducted among discerning guests of luxurious hotels to gauge the factors that would make them more devoted to a particular establishment. The survey involved detailed interviews, which helped create a list of 18 potential amenities. The guests were then requested to rate each feature on a scale of 1 to 7, with 1 denoting no impact on loyalty and 7 indicating an enormous effect on loyalty. In a separate section of the survey, guests were asked to identify the amenities currently offered at the hotels they were loyal to. By comparing the amenities, the hotels offer with what the guests desire, one can easily identify areas where hotels are either meeting or falling short of their guests' needs. This comparison is commonly referred to as a gap in performance versus importance.

Service Quality

The interpretation of quality can vary from person to person. Parasuraman et al. (1985) describe quality as "the level to which a customer perceives the service to meet or exceed their expectations." Thus, service quality is determined by comparing a customer's perception of the service they received to their desired level of service. The service is considered exceptional and a pleasant surprise if their expectations are surpassed. However, the service is viewed as unacceptable if expectations are not met. Customers become content when their expectations are satisfied. According to Ali et al. (2021), the hospitality industry should prioritise fulfilling its customers' needs, requirements, and expectations to enhance the company's reputation. This is because the service industry faces unique challenges compared to product-based industries due to the nature of the service provided.

Customer Satisfaction

The degree of a customer's satisfaction with a purchase is contingent upon the product's ability to meet, exceed, or fall short of their expectations. Dissatisfaction may arise if the product does not meet their expectations, while satisfaction is achieved if it meets their expectations. Additionally, a customer may experience high levels of satisfaction or even delight if the product surpasses their expectations.

Buyers tend to form their expectations based on their past buying experiences, the opinions of their peers and colleagues, and information from marketers and competitors. As marketers, we are responsible for setting expectations carefully to attract buyers without disappointing them. Setting expectations too low may satisfy some buyers, but it may fail to impress others. Conversely, setting expectations too high can lead to disappointment. For instance, a campaign by Holiday Inn called "No Surprises" promised consistently trouble-free accommodations and service but failed to meet customers' expectations, leading to dissatisfaction. It is essential to understand that customer satisfaction with service quality mediates between customer experience and loyalty (Lemy et al., 2019; Priporas et al., 2017).

Relationship Marketing

To effectively cultivate customer loyalty, it is recommended that managers focus on establishing strong and enduring relationships through the implementation of relationship-marketing strategies. This approach involves nurturing connections with customers and other stakeholders who can contribute to providing exceptional service. Rather than concentrating solely on individual transactions, marketing efforts are shifting towards developing long-term relationships and networks. The ultimate aim is to provide lasting value to customers and ensure their sustained satisfaction. To achieve these goals, it is essential that all departments within the organisation work collaboratively as a team to provide the best possible service to the customer. Relationship marketing involves establishing connections on multiple levels, including economic, social, technical, and legal, ultimately leading to elevated levels of customer loyalty.

Relationship Marketing vs. Traditional Marketing

Marketing delivery travels along a continuum from basic traditional delivery through a relationship status where profitable partnerships are formed. The steps are outlined next:

- *Basic*: The company sells the product, but unfortunately, they do not provide any follow-up services.
- *Reactive*: The company sells the product and offers customer support for any questions or issues. They encourage customers to reach out for assistance at any time.
- *Accountable*: The customer can expect a prompt phone call from a company representative after booking. During this conversation, the representative will happily address any inquiries or concerns the customer may have. Additionally, the salesperson will kindly request any constructive feedback regarding potential areas of improvement, or any specific disappointments experienced during or after the event. This valuable feedback will enable the company to continuously enhance its offerings for future customers.
- *Proactive*: Occasionally, the company's representatives may initiate contact with the customer to offer helpful suggestions on how to elevate their experience while also presenting creative ideas for future events.
- *Partnership*: Our organisation is constantly committed to enhancing the quality of service we offer our valued clientele. We achieve this by establishing a close working relationship with our clients and collaborating with other customers to uncover novel and effective solutions. By working together, we can create and maintain a culture of innovation and excellence that benefits everyone involved.

To improve customer satisfaction and cultivate stronger relationships, what marketing strategies can a company employ? There are three approaches to create customer value. The first involves providing financial incentives through loyalty programmes. For instance, airlines may implement frequent-flyer programmes, hotels could offer room upgrades to regular guests, and restaurants may establish frequent dining programmes. While such reward initiatives and financial incentives can foster customer loyalty, competitors can easily duplicate them. For example, Marriott offers three levels: gold (15 nights), black (50 nights), and platinum (75 nights), while Hilton offers silver (10 nights), gold (36 nights), and diamond (60 nights). As customers progress to higher tiers, they receive additional benefits.

Rather than exclusively focusing on financial gains, an alternative approach is to incorporate social benefits as well. This entails personnel within a company taking the time to fully comprehend each customer's unique needs and desires and crafting their products and services accordingly. By doing so, customers are elevated to the status of clients, enjoying customised and personalised experiences. These clients are serviced on an individual level by professionals who are devoted to their specific needs rather than being served by just any available staff member. For example, a server might greet regular patrons by name, while a salesperson may develop a close relationship with their clients. These social connections encourage clients to return and may even result in clients following the staff member with whom they have bonded if they move to a new workplace. Managers in the hospitality and travel industries ensure that significant clients develop social bonds with multiple individuals within the organisation. This includes the general manager, front desk manager, food and beverage manager, convention services manager, banquet manager, and restaurant manager. The general manager may also accompany sales personnel on calls to important clients. In this way, clients feel as though they have established relationships with key individuals within the organisation rather than relying solely on the salesperson.

Third, to foster strong relationships with their customers, businesses can establish tangible connections and offer financial and social perks. Airlines, for example, have implemented booking systems specifically for travel agents and designated phone lines for frequent flyers. They have also created upscale lounges for their first-class clientele and extended limousine services to the airport. Similarly, Sheraton provides flexible check-in and checkout times for their most valued guests, while Hilton utilises advanced technology to deliver personalised welcome messages on guests' in-room televisions. Furthermore, Hilton is exploring the potential of a customised tracking system that would provide guests with a proximity card, allowing staff to locate them on the premises. The company is also developing "Wireless Anticipated Information," which would enable messages from the hotel to be forwarded to guests' personal digital assistants. While implementing such structural changes can be challenging, they offer a distinct competitive advantage that is difficult for rivals to replicate. To establish a relationship marketing programme in a company, follow these main steps:

- Determine which customers are most important for relationship management. Focus on the biggest or most valuable customers and assign them to relationship management. You can also consider adding customers who demonstrate significant growth or are innovating in their industry.
- In order to enhance the level of customer satisfaction, it is highly recommended that a professional relationship manager be assigned to

every significant customer. If the current salesperson lacks the necessary skills in relationship management, it would be advisable to provide them with the appropriate training or opt for a replacement with greater suitability for the position. The relationship manager should possess qualities that are in line with, or that resonate with, the customer.

- To ensure the smooth handling of client relationships, it is of utmost importance to create a detailed job description that clearly defines the reporting relationships, objectives, responsibilities, and evaluation criteria for all relationship managers. It is highly recommended that each relationship manager is assigned a specific number of client relationships to manage so that they can focus on delivering personalised service and serve as the main point of contact for all client interactions.
- It is recommended that relationship managers consider creating customer relationship plans on an annual basis and for the foreseeable future. These plans should cover objectives, approaches, practical measures, and vital resources.
- To increase the effectiveness of our relationship managers, it would be beneficial to appoint an overall manager to supervise them. This person can develop job descriptions, and evaluation criteria, and provide resource support to ensure the relationship managers are performing at their best.
- After successfully implementing relationship management, the organisation can allocate resources towards managing its customers and products. Nevertheless, it is worth noting that although many businesses are adopting relationship marketing, it may not always yield the desired results in every circumstance.

Relationship marketing has become increasingly important as companies strive to build long-term connections with their customers. However, not all customers are created equal, and it is crucial for companies to consider which customers are deserving of a cultivated relationship carefully. This requires a thoughtful evaluation of each customer's needs, preferences, and potential value to the company. By focusing their efforts on those individuals who can truly benefit from their services, companies can build stronger, more meaningful relationships that lead to increased loyalty, repeat business, and, ultimately, greater profits. Therefore, companies must invest the time and resources necessary to identify and nurture the right customers while avoiding the temptation to spread themselves too thin by trying to build relationships with everyone.

Retaining Customers

Relationship marketing can provide various advantages, such as establishing a loyal customer base that remains committed to your business, minimising marketing expenses, reducing price sensitivity among loyal customers, and enabling loyal customers to participate in partnership activities. By retaining loyal customers, your business can save on marketing costs, as it requires fewer resources to maintain a customer than to acquire a new one. Furthermore, loyal customers are more inclined to recommend your business to others, which can result in positive word-of-mouth marketing. They are also less likely to switch to a competitor due to price and tend to make more purchases than disloyal customers. Finally, partnership activities of loyal customers can include a range of activities, such as business referrals, providing references, publicity, and serving on advisory boards. The advantages of having a loyal customer base can be significant, and even a small increase in loyal customers can lead to a substantial increase in profitability. According to a study by Reichheld and Sasser, a 5 percent increase in customer retention resulted in a 25 percent to 125 percent increase in profits across nine service industry groups they studied. Therefore, it is imperative for service firms to prioritise building strong relationships with their customers.

In recent years, the market has become more competitive, with many products reaching a state of maturity. As a result, it can be challenging to differentiate between products in the same category. For example, when presented with images of hotel rooms from Sheraton and three other competitors, managers in Asia had difficulty identifying one of the rooms' brand identities, despite being provided with a list of eight brands to consider. This exercise illustrates the difficulty of distinguishing between hotel brands based on physical attributes. This intensified competition, with a lack of product differentiation, led to the emergence of relationship marketing in the 1990s. Relationship marketing aims to foster customer loyalty by treating them as partners, in contrast to traditional marketing, which is often more transactional.

In the current business landscape, customer satisfaction plays a crucial role in determining the success of a company. To ensure that customers remain loyal, businesses must prioritise delivering high-quality products and services. Gone are the days when companies could take their customers for granted, assuming that they had no other options. In today's fiercely competitive environment, retaining existing customers is just as important as acquiring new ones. By investing in building stronger relationships with their customers, companies can save money on customer acquisition costs and boost their overall revenue. It's essential to remember that customer loyalty is the key to sustained business growth.

As a business, receiving negative feedback from dissatisfied customers can be a tough hurdle to overcome. It can be especially challenging when your primary focus is on attracting new customers regularly. For instance, if you own a local restaurant, losing customers due to negative word of mouth can be disheartening. However, it's essential to prioritise addressing customer concerns and working to deliver positive experiences to help minimise the impact of negative feedback. By doing so, you can create a more welcoming and positive environment for your customers and improve your business's overall reputation.

Cost of Lost Customers

Reducing customer defection rates is critical for companies, and measuring the retention rate is the first step to achieving this. Once the reasons for customer defection are identified, the company can take necessary actions to reduce or eliminate them. Although some factors like customers moving or going out of business are beyond control, poor service, low-quality food, and high prices can be addressed comprehensively. Creating a frequency distribution that shows the percentage of customers who defect for different reasons is vital in this regard.

Estimating the profit lost from unnecessary customer defection is equally important. The loss is equivalent to the customer's lifetime value, which is the total income expected from the customer throughout their relationship with the business. For instance, Ritz-Carlton values its repeat customers at over $100,000 throughout their lifetime. This highlights the significance of retaining customers and how it can result in a substantial monetary benefit for the company.

Businesses understand the immense value that a single customer can bring. Whether it's a restaurant patron or a travel agency client, their lifetime value can amount to thousands of dollars. To determine this value, businesses take into account the average amount a customer spends per year and multiply it by the length of time they remain a customer. This length of time can be affected by various factors such as job changes and location. For instance, a business traveller might only remain a customer for roughly four years, while a restaurant customer in a transient community may only last around three years. However, with a Customer Relationship Management (CRM) system in place, hotel chains can easily track their guests' lifetime value across all locations and market segments.

In today's competitive market, it is essential for companies to maintain their existing customer base rather than solely focusing on acquiring new customers. This strategy is more cost-effective, as attracting new customers can be considerably more expensive than nurturing customer satisfaction. Therefore, it is advisable for companies to consider investing in measures that can minimise customer defections if the expenses incurred are lower than the losses incurred from losing customers. It is noteworthy that defensive marketing, which prioritises keeping current customers satisfied, is typically more economical than offensive marketing, which aims to draw customers away from competitors.

Historically, businesses have placed more emphasis on attracting new customers than cultivating lasting relationships with existing ones. This approach prioritised transactions over loyalty. However, contemporary companies recognise the value of retaining their current clientele. Studies reveal that reducing customer churn by just 5 percent can increase profits by up to 85 percent. Unfortunately, most accounting systems fail to accurately assess the true worth of loyal customers.

Customer retention should be a top priority for companies, as it is the key to building customer loyalty. While marketing efforts may focus on acquiring new customers and generating sales, excellent customer satisfaction is essential for retaining customers and ensuring their continued support.

The Link Between Marketing and Quality

According to a recent study conducted by PricewaterhouseCoopers, the top 50 firms in America prioritise quality and customer service. This applies particularly to the hospitality industry, which has long recognised the importance of providing high-quality service to ensure customer satisfaction and repeat business. Articles focusing on quality in this industry have seen a significant surge in popularity since the late 1980s. Moreover, Ritz-Carlton's historic achievement in 1992, when it became the first hospitality company to receive the prestigious Malcolm Baldrige National Quality Award, further reinforced the growing interest of hospitality firms in service quality.

There exist varying viewpoints regarding the definition of quality. While some, such as Philip Crosby, maintain that it is the adherence to established company standards, others argue that it is gauged by the ability to meet or surpass customer expectations. Furthermore, some consider the transition from a two-star to a four-star hotel as a sign of improved quality. However, these definitions may prompt more inquiries than answers. Is quality truly free, as Crosby asserts, or does it come at a price? To clarify this concept, we shall define quality, review quality models, examine its connection to marketing, stress its significance, and suggest strategies for enhancing product quality within the hospitality industry.

In the hospitality industry, employees and guests interact frequently, making coordination crucial. Achieving complete quality is impossible, as mistakes are inevitable and systems may fail. However, every hospitality organisation must strive for continuous improvement through total quality programmes. Managers can enhance guests' perception of their experience by eliminating failures and improving product quality. Companies that neglect quality standards may face significant financial consequences.

Upon the arrival of a recent food and beverage manager at the Ohio University's University Centre, an unfortunate situation arose. The food-service workers went on strike, leaving management with no choice but to quickly train unskilled students to fill all the positions in the operation. The timing of the strike coincided with the centre's ongoing efforts to sell its facilities for the evening dinner meeting of the Rotary Club, making the situation even more challenging. Despite the obstacles, the Centre remained committed to providing exceptional service to all of its guests, and the students stepped up to the occasion with grace and professionalism. The newly appointed food and beverage manager was acutely aware of the importance of the upcoming occasion. To ensure that the event would be a resounding success, the manager made the decision to create a customised menu that would truly impress the attendees. One of the standout dishes was a delectable beef stroganoff, a dish that the manager had carefully crafted using only the finest beef tenderloin. Interestingly, the inspiration for this dish came from the manager's time spent in college. During a particularly memorable food class, the students had the privilege of watching a well-known chef conduct a demonstration. The chef's beef stroganoff had received high praise from the students, but the instructor had dismissed the chef's abilities, claiming that anyone could make a delicious beef stroganoff with tenderloin.

The manager took careful notice of the situation and set out on a mission to craft his own rendition of the dish, utilising the same premium ingredients. The outcome was a delectable creation bound to leave a lasting impression on anyone fortunate enough to taste it. The sauce was exceptional, and the salads were artfully presented. The manager aspired to establish a regular patronage from the Rotary Club. However, during the event, the manager observed that many guests were leaving a substantial portion of their beef stroganoff uneaten. It was then discovered that a bag of stew meal cubes had been mistakenly used instead of beef tenderloin cubes. This oversight led to the university center losing the group, which could have generated $10,000 annually, or $50,000 over five years. This error underscores the importance of quality control in maintaining customer satisfaction.

What Is Quality?

Service quality is often defined as how well a customer's perception of service meets or exceeds their expectations. There are two types of quality: product features and freedom from deficiencies. Product features can increase customer satisfaction, but they also raise the cost of the product. Customers must be willing to pay for these additional features or be more loyal to the brand. For instance, McDonald's charges more for hamburgers that come with lettuce and tomato. Similarly, hotel rooms on concierge floors have more amenities than regular ones and can command higher prices. La Quinta Inns offers free local phone calls to foster loyalty among salespeople. Ultimately, quality is subjective and varies from person to person.

Benefits of Service Quality

In the previous part of this chapter, we explored the advantages of customer loyalty. Now, we delve deeper into the benefits that stem from customer loyalty resulting from superior quality. Companies with a larger market share and a superior perceived quality compared to their competitors can achieve significantly higher returns compared to those with a smaller market share and lower quality. In *The PIMS Principles* book, the authors establish a connection between quality and profitability, exemplified in Figure 11.3. As depicted in the figure, firms with both high market share and high quality experience the most substantial return on investment.

Customer Retention

Providing exceptional service is paramount in establishing a devoted customer base and fostering positive recommendations. It plays a significant role in the customer's purchasing decision and impacts their overall satisfaction, which in turn influences their likelihood of returning and referring others. According to studies, it is much more expensive to acquire a new customer than to maintain an existing one, with some estimates suggesting that it is up to four to six times more expensive. Bill Marriott famously noted that it costs $10 to get a guest to stay at a Marriott hotel for the first time but only $1 to persuade them to return. Once a customer is content with a particular hotel, it is difficult to entice them to switch to a competitor, even if the latter is providing a significant discount. Salespeople may have to wait for a mistake from a rival hotel before they can convince a customer to try their hotel, which could take months or even years. During this period, salespeople will be making calls, distributing advertising materials, and inviting potential clients to breakfast or lunch at the hotel. The hotel will be spending money on advertising, PR, and direct-mail campaigns to attract potential clients, which can cost thousands of dollars. If a significant client decides to use the hotel, the money spent on marketing is well spent. However, if a potential client has a negative experience with the hotel's service quality, they are likely to leave. In such cases, all marketing efforts aimed at acquiring this customer will have been in vain.

A happy customer is likely to recommend your business to others through word-of-mouth. On average, a satisfied guest will tell about five people, while an unhappy customer may tell ten or more people or post negative feedback on social media. To maintain a positive reputation, at least two or more customers should leave feeling satisfied for every dissatisfied customer. If a hotel or restaurant receives mixed reviews, it may be perceived as mediocre by the market. Therefore, businesses striving for a strong reputation must aim for consistently high-quality service.

Achieving excellence in the hospitality industry requires minimising errors. A 200-room hotel accommodates over 50,000 guests annually, and a standard adherence rate of 90 percent is typically acceptable. However, if housekeepers only adhere to the hotel's specifications 90 percent of the time, around 5,000 guests may receive rooms that do not meet the hotel's standards. It's important to note that even if some guests may not notice any difference, poor quality can significantly affect financial losses. For instance, if 2,500 guests notice the difference and 1,250 of them choose not to return, the hotel may lose up to $125,000 in future revenue. When considering other areas such as food and beverage, front desk, and reservations, it's clear that maintaining high quality is crucial for achieving financial success.

Avoidance of Price Competition

As suggested by the renowned chicken producer Frank Perdue, customers are often willing to pay a premium for superior products, even if it requires additional effort on their part. According to PIMS data, companies that prioritise quality may charge a slightly higher price, approximately 5–6 percent, than those that do not. Focusing on quality can reduce the need for competitive pricing and enhance revenue potential.

A restaurant that is known for its high-quality food and excellent service has a significant advantage over one that is known for inconsistent or poor quality. Such a restaurant can rely on positive word-of-mouth and customer recommendations to attract new business. On the other hand, a restaurant with a poor reputation will struggle to retain repeat customers and may experience more negative word-of-mouth than positive. In such cases, restaurants often resort to price discounts, such as two-for-one coupons, to attract customers.

Unfortunately, some hospitality companies fail to focus on understanding their customers' true preferences. For certain hotel guests, having newspapers available in their rooms might be more valuable than purchasing an expensive gym membership. In the hospitality industry, differentiating products can be as simple as directly asking guests about their preferences and needs. By doing so, companies can better cater to their customers' desires and enhance their overall experience.

Retention of Good Employees

It is widely acknowledged that employees appreciate well-organised operations that yield superior products. It has been noted that front-desk clerks are not fond of receiving complaints from guests. The unfortunate consequence of poor quality can be absenteeism, increased turnover, and lower morale among employees, which can lead to additional costs. According to recent hotel and restaurant management school graduates, lack of quality is cited as a reason for leaving their jobs. Conversely, high-quality fosters employee loyalty, facilitates recruitment efforts, and helps to minimise training costs.

Cost Reduction

There are three types of costs associated with maintaining quality standards: internal costs, external costs, and quality system costs. Internal costs are incurred by the company when they need to address issues with the product before it reaches the customer. An example of this would be if an air conditioner malfunctions due to poor maintenance, which could lead to the unavailability of the guest room until the issue is resolved. Similarly, if a cook mistakenly prepares fried grouper instead of grilled grouper, the server will identify the error and have the cook prepare a new piece of fish.

In the hospitality industry, it can be difficult to identify errors before they affect the customer experience. When a customer's expectations are not met, it may lead to additional costs for the organisation in resolving the issue. These costs can be significant if the customer decides to not return due to poor service. To mitigate these external costs, organisations have provided gestures of goodwill such as a complimentary bottle of wine, breakfast, or fruit basket. It is important for organisations to make a concerted effort to recover from service failures and address customer complaints in a timely and effective manner.

The provision of a high-quality service system is a crucial aspect of any business. While it may entail some costs, these expenses are typically lower than those resulting from poor-quality service. It is essential to invest in quality system costs, such as customer service audits, training, management meetings, and new technology implementation, as these ensure that the company remains competitive and retains customers.

In contrast, internal costs that do not affect customer satisfaction are considered a waste of money. These expenses should be minimised wherever possible to ensure that the company's resources are utilised effectively. On the other hand, external costs related to service failures can be significant. In such instances, companies may need to spend a considerable amount of money to maintain customer goodwill and prevent the loss of customers.

However, it is important to note that successful service recovery can lead to higher customer satisfaction and loyalty. By addressing any issues that may arise promptly and effectively, companies can turn negative experiences into positive ones and enhance their reputation. Despite these efforts, some customers may still choose to leave the company. It is crucial to acknowledge this possibility and work towards minimising the number of customers lost. Overall, investing in a high-quality service system is a wise decision that can lead to long-term success and growth for any business.

Summary

In this chapter, we clarified the distinction between customer satisfaction and customer loyalty. Additionally, we delved into the concept of relationship marketing as a means of both acquiring new customers and retaining current ones.

Activity

1. Define quality and discuss its importance in the hospitality industry.
2. How can managers develop and sustain good customer relationships in the hospitality industry?
3. Explain the effects of customer loyalty and customer satisfaction on performance in the tourism industry in Africa.
4. Define customer value and customer satisfaction and discuss any two benefits of each.
5. Distinguish between customer satisfaction and customer loyalty.
6. Discuss any three ways through which managers in the hospitality industry can attract new customers and retain current customers through relationship marketing.
7. Discuss any three reasons why complaint management in the hospitality industry is crucial.
8. Explain any three strategies that could be employed in the H&T industry for resolving customer complaints.
9. Define quality and discuss its importance to tourism and hospitality businesses.

References

Ali, B. J., Gardi, B., Jabbar Othman, B., Ali Ahmed, S., Burhan Ismael, N., Abdalla Hamza, P., ... Anwar, G. (2021). Hotel service quality: The impact of service quality on customer satisfaction in hospitality. *Engineering, Business and Management*, *5*(3), 14–28.

Lemy, D., Goh, E., & Ferry, J. (2019). Moving out of the silo: How service quality innovations can develop customer loyalty in Indonesia's hotels. *Journal of Vacation Marketing, 25*(4), 462–479. https://doi.org/10.1177/135676671881965

Nobar, H. B. K., & Rostamzadeh, R. (2018). The impact of customer satisfaction, customer experience and customer loyalty on brand power: Empirical evidence from hotel industry. *Journal of Business Economics and Management, 19*(2), 417–430.

Parasuraman, A., Zeithaml, V. A., & Berry, L. L. (1985). A conceptual model of service quality and its implications for future research. *Journal of Marketing, 49*(4), 41–50.

Priporas, C. V., Stylos, N., Vedanthachari, L. N., & Santiwatana, P. (2017). Service quality, satisfaction, and customer loyalty in Airbnb accommodation in Thailand. *International Journal of Tourism Research, 19*(6), 693–704. https://doi.org/10.1002/jtr.2141

Chapter 23

Electronic Marketing: Internet Marketing, Database Marketing, and Direct Marketing

Chapter Outline

- Introduction
- Electronic Marketing
- What Is Electronic Marketing?
- Internet Marketing
- What then Is the Internet?
- Interactivity and Accessibility of the Internet
- Internet Marketing through Websites
- Making the Website Attractive
- Database Marketing
- Major Principles of Database Marketing
- Direct Marketing
- Reasons for Growth of Direct Marketing
- Development of Integrated Direct Marketing

DOI: 10.4324/9781032688497-23

Objectives

At the end of this chapter, students should be able to:

- Define and explain electronic marketing.
- Identify and explain concepts under electronic marketing.
- Identify and explain the steps involved in creating effective websites.

Introduction

E-marketing, also known as electronic marketing, refers to the utilisation of information and communication technology (ICT) to create, communicate, and deliver value to customers, clients, and partners. Its purpose is to foster beneficial relationships that benefit the organisation and its stakeholders. As information and communication technology continues to advance, e-marketing is gaining significance in the field of marketing, and there is a growing belief that it may become the primary approach to conducting marketing activities. This section aims to elucidate the concepts of e-marketing and its relevance, showcasing how it can be effectively employed to benefit the hospitality and tourism industry.

Electronic Marketing

The Hospitality and Tourism sector, much like other industries, operates within a dynamic and ever-evolving environment, requiring constant monitoring and adaptation by all industry participants. Presently, the industry confronts a range of emerging challenges stemming from shifts in customer behaviour and environmental conditions (Middleton et al., 2009). These challenges encompass:

- There is a requirement to streamline the service offerings provided by the industries to consumers, while maintaining the necessary flexibility to adapt to shifts in the operational environment.
- There is a necessity to create and effectively deliver products and experiences that align with the diverse demands of different markets. Accomplishing these objectives necessitates the adoption of a strategic marketing management approach, alongside the availability of accurate and dependable information.

The swift progress of information and communication technologies (ICTs), particularly the internet and the world wide web, has presented both challenges and prospects for industries across the board, including the hospitality and tourism sectors. ICTs and the internet have significantly transformed the manner in which companies execute numerous business processes, with marketing being notably impacted due to the opportunities arising from online communications (Jensen, 2006). It is evident that ICTs have a substantial influence on nearly every facet of marketing, including the key stages in the marketing mix (Kotler, 2003; Siegel, 2003).

The impact of ICT and the internet on marketing has led to the emergence and development of electronic marketing (e-marketing). The significance of e-marketing in modern marketing practices is so profound that it is anticipated to become the prevailing method of conducting marketing activities in the near future.

What Is Electronic Marketing?

Electronic marketing, also known as e-marketing, is the utilisation of information technology for marketing activities. It involves the use of information technology (ICT) to create, communicate, and deliver value to customers, clients, and partners, while also managing relationships for the benefit of the organisation and its stakeholders. E-marketing is a subset of a broader category called e-business, which encompasses the use of electronic platforms such as intranets, extranets, and the internet to conduct a company's business. E-commerce, another component of e-business, specifically focuses on using electronic platforms for transactional purposes. While e-marketing primarily refers to the application of information technology to traditional marketing practices, it is often used interchangeably with internet marketing. However, in this chapter, various aspects of electronic marketing, including internet marketing, database marketing, and direct marketing, will be discussed, highlighting how their advantages can be leveraged for effective management of hospitality and tourism marketing.

Internet Marketing

The development of the world ide web and web browsers during the 1990s has elevated the internet from a simple communication tool to a significant technology that has revolutionised various aspects of our lives. With the increasing availability of technology devices like computers and internet-enabled mobile phones, internet accessibility has expanded globally, transforming the internet into a powerful tool for conducting business operations, including marketing. In the hospitality and tourism industries, harnessing the internet for marketing has become crucial for gaining a competitive edge and has, in many ways, revolutionised traditional marketing practices.

The widespread accessibility of the internet and the growing availability of computers and connectivity have made it increasingly convenient for hotels of any size to engage in electronic communication with other businesses and consumers. The internet has fundamentally transformed the dissemination of tourism information and sales, granting tourism destinations and businesses direct access to end consumers. Simultaneously, it has become the primary channel for business-to-business (B2B) communication. In essence, there are numerous opportunities for enhanced business-to-business (B2B) communication and direct selling to consumers (B2C). Furthermore, the digital marketplace facilitated by the internet has facilitated the rapid emergence of new products and services, leading to reduced marketing costs, the elimination of intermediaries, and a redefinition of marketing relationships (Strauss et al., 2014).

What Then Is the Internet?

While you may already have some familiarity with using the internet in your daily life, have you ever taken a moment to ponder what the internet truly entails? It would be beneficial for us to briefly contemplate its nature.

The internet is a vast network that connects millions of computers worldwide. It allows these computers to communicate with each other and share information. Think of it as a global system where computers can send and receive data using special rules called the Internet Protocol. This data can be transmitted through phone lines, cables, and satellites, enabling people to access and exchange all sorts of content, like text, images, and videos, from anywhere in the world.

The internet is made up of three different types of networks: the intranet, extranet, and web. An intranet is a network used within a company or organisation, following the same standards as the internet. It allows for internal communication and information sharing. An extranet is an extension of the intranet that connects the organisation with its partners, such as suppliers or distributors, for collaborative purposes. On the other hand, the web is the part of the internet that we commonly interact with. It provides a user-friendly interface with features like text, images, sound, and videos. It enables us to navigate through different web pages using a browser and allows for real-time interaction, discussions, and feedback. The web is the main platform where users engage with the internet's content and services.

Interactivity and Accessibility of the Internet

The internet and the web possess unique attributes that offer opportunities that surpass those provided by traditional communication media such as telephone, television, and postal mail. The interactive nature and widespread accessibility of the internet enable more effective and efficient marketing strategies and tactical implementations. Furthermore, these properties of the internet have fundamentally altered the methods and approaches used in marketing within the hospitality and tourism sector (Lustria, 2007).

In the context of the hospitality and tourism industries, the internet serves as a comprehensive platform for accessing diverse forms of information. This information encompasses various formats such as audio, video, virtual tours, text, images, and even digital transactions, all without any geographical limitations. The range of information available includes details about hotel facilities, room types, services offered, reservation and pricing information, local area highlights, company information, and affiliations (Lustria, 2007). More recently, websites have enabled online reservations with instant confirmation, allowing guests to conveniently book hotels through the internet. Additionally, hotels can easily manage rates and availability across multiple channels using a user-friendly interface. This level of flexibility and control can be accessed at any time and from any location with internet connectivity.

Potential guests and tourists have multiple options for accessing the internet, including PCs, web TVs, mobile phones, and other devices. Access to pre-trip, in-trip, and post-trip information is crucial for travellers worldwide. Pre-trip information search involves planning aspects such as transportation, accommodation, attractions, and weather conditions. Internet access is vital for gathering information and facilitating transactions. The internet can be accessed through various devices, including TVs, mobile devices, and in-car technology. Given the decreasing cost and increasing speed of internet access, players in the hospitality and tourism industry should prioritise efforts in this area.

Internet Marketing through Websites

Developing a Website

When the core concepts of internet marketing are carefully analysed, targeted promotional strategies suitable for hospitality and tourism businesses can be identified. One effective approach is to prioritise the establishment of an independent website as the initial step. This strategy is not only cost-efficient but also critical for the success of hospitality and tourism enterprises. By creating a dedicated website, these businesses can successfully attract potential customers and guide them to a centralised online platform where they can access essential information and make well-informed decisions about the offered services.

A hospitality or tourism company has the option to create various types of websites, but the fundamental one is a corporate website. These websites are primarily intended to establish a positive relationship with customers and support other sales channels, rather than directly selling products. In contrast, marketing websites aim to actively engage consumers in interactions that drive them towards making a direct purchase or achieving other marketing objectives (Kim, 2020).

There are several affordable methods to establish an independent online presence, ranging from free hosting services, free weblog providers (e.g., www.wordpress.com and Google's Blogspot www.blogspot.com) to self-built resources such as CNET's or professional services offered by companies specialising in the tourism and hospitality sector. While the latter option may come with higher costs, it remains a highly cost-effective choice for most hospitality and tourism businesses. When creating an effective website, it is crucial to consider various important factors (Bala & Verma, 2018). These include:

1. Clearly defining the website's objectives.
2. Identifying the target audience.
3. Understanding their readiness to make a purchase.
4. Developing concise and persuasive messages.
5. Addressing frequently asked questions, avoiding unnecessary technological gimmicks.
6. Actively promoting the website.

Defining the Website's Objectives

There are several factors influencing businesses to develop and maintain their independent websites. For many businesses, the primary purpose of establishing an online presence is to promote the organisation's name, convey its purpose, and deliver information for both existing potential customers. In the case of hospitality and tourism enterprises, the objectives for developing a website may differ to some extent, typically encompassing multiple goals. The main objective for these businesses is often to generate more business by attracting new customers. Additionally, hospitality and tourism enterprises may use their website to facilitate online transactions, gain insights into customer preferences, share details about special offers, or support sales inquiries that occur offline (Giao et al., 2020). It is crucial to clearly define the objectives for creating a website before proceeding with its development.

Clearly Identifying the Target Market

Like any other form of marketing communication, the creation of a successful independent website begins with a clearly defined target audience. In order to identify this target audience, it is important for hospitality and tourism enterprises to conduct an analysis of both their current customer base and potential markets. This analysis provides valuable insights that inform the development of the website, guiding decisions about the content to be included and how it should be presented. By understanding their audience, hospitality and tourism enterprises can tailor their website to effectively communicate with and cater to the needs of their target market.

Establishing the Buyer-readiness Stage

After establishing a clear understanding of the target audience, it is important to consider the stage of the buyer-readiness sequence that the audience has progressed to.

Typically, visitors to an independent website are already familiar with the business and have some level of interest in its products or services. As a result, the design of the website should cater to the needs of potential customers who have moved beyond the initial "awareness" stage. These individuals are now seeking more comprehensive information about the offerings provided by the business. Therefore, the website should be structured and designed in a way that facilitates the gathering of detailed information, ensuring it meets the needs of these potential customers who are actively seeking more in-depth insights.

Developing Simple but Persuasive Messages

The independent website should present concise and compelling messages. It should have a clean and fast-loading design. The details of the company such as email address should be easily accessible. Specific information such as room configurations or tour itineraries should be provided on separate pages and linked within relevant sentences, allowing visitors to click and access further details. Similarly, large images that may slow down loading times can be presented as descriptive text links or captioned thumbnail images, giving visitors the choice to prioritise loading time based on the perceived importance of the image. The content of the website should be persuasive and engaging, encouraging visitors to take action by making a purchase or initiating further inquiries.

Answering the Most Frequently Asked Questions

An independent website should offer introductory details regarding the products or services offered by the business. It should function as a self-contained platform where prospective customers can find all the necessary information about the business. The information presented on the website should be thorough and, at a minimum, address the most commonly asked questions, such as the business's location, operating hours, and pricing details.

Avoiding the Use of Excess Graphics

When creating a website, it is advisable for businesses to refrain from incorporating excessive graphics that offer minimal additional information about the business and do not impact purchasing decisions. Still images and text can often convey information just as effectively as streaming audio and video. Additionally, it is best to avoid displaying visitor counters on the website. Low visitor numbers may discourage visitors, while high numbers can undermine the credibility of the site.

Promoting the Website

Once the website is finalised, the next step is to attract visitors to it. The site's address or URL should be prominently displayed on all promotional materials. Additionally, the website can be utilised as a cost-effective tool to complement traditional marketing campaigns. For instance, instead of printing numerous entry forms for a competition, interested participants can be directed to an online entry form. While there are various methods to promote a website and boost visitor numbers, one of the most cost-effective approaches is to establish links from other websites, including specialist and mainstream search engines and directories. Expanding the distribution of the site is of utmost significance in attracting visitors.

Making the Website Attractive

The key to attracting and retaining consumers on a website is to offer sufficient value and excitement that encourages them to visit, stay engaged, and return in the future (Mohseni et al., 2018). The following are key areas H&T practitioners should concentrate on to build attractive websites:

- *Content*: To entice customers to revisit the site, it is crucial to provide valuable and relevant content. The website should be well-organised, allowing users to easily access the information they seek while projecting an image that aligns with the product or brand.
- *Security*: In the case of marketing websites where purchases are made, customers are drawn to websites that demonstrate robust security measures to protect their personal information, especially payment-related details. This builds trust and confidence in the website's reliability.
- *Legal and Ethical Considerations*: Companies have a legal and ethical obligation to inform customers about the use of their information and to prioritise customer privacy while adhering to relevant privacy laws and regulations. Respecting customer privacy and maintaining transparency regarding data usage are essential.

Overall, by providing valuable content, ensuring secure transactions, and adhering to legal and ethical guidelines, businesses can create an attractive website that not only captures consumer interest but also builds trust and fosters long-term customer relationships.

Database Marketing

A marketing database is a well-organised compilation of data regarding individual customers, prospects, or potential customers. It is designed to be easily accessible and actionable for various marketing purposes, such as lead generation, lead qualification, sales, or maintaining customer relationships. The primary objective of database marketing is to provide the company with a comprehensive dataset of information about the desired stakeholders, typically the customers, which aids in developing predictive models. These predictive models are then utilised to anticipate how stakeholders are likely to respond to specific company actions. This valuable information enables the company to deliver targeted messages at the right time, through appropriate channels, to the right individuals. As a result, customers are satisfied, response rates per marketing investment increase, the cost per order decreases, and overall profits are enhanced.

The integration of the internet and database marketing has become a standard practice for most companies. While database marketing employs various direct marketing techniques such as response advertising, direct mail, and telemarketing, it adopts a comprehensive marketing approach that emphasises interactive communication with customers, rather than one-way communication.

Major Principles of Database Marketing

Although the following statements may appear evident, it is essential to reiterate them as they emphasise the significance of the customer relationship:

- Previous consumer behaviour provides the most accurate indication of future behaviour.
- A purchase represents only one instance in a customer's lifetime. To accurately assess a customer's value, it is necessary to calculate their lifetime value.
- Existing customers hold greater importance than potential customers.
- Certain customers hold greater significance than others.
- Customers are more likely to possess specific characteristics that are worth considering.

These principles reflect the well-known 80/20 rule, which states that 80 percent of a business's revenue is generated by 20 percent of its customers. It is also recognised that retaining existing customers is more cost-effective than acquiring new ones. Following these principles necessitates the establishment of an efficient database and places significant demands on an organisation. Every interaction with customers must be meticulously recorded to develop a comprehensive understanding of their purchasing behaviour and value. Customers should be ranked or assessed based on their importance in the database. The user must decide which data to include in the database and determine how to analyse and utilise the information in the marketing planning and implementation process.

Data Warehousing

At an advanced stage, a customer database can serve as a valuable tool for forecasting future customer behaviour. However, this necessitates access to a comprehensive range of information. Larger travel and tourism organisations are increasingly considering the establishment of a data warehouse to facilitate such predictions. The primary goal of a data warehouse is to consolidate customer data from various departments of the organisation, allowing for a more comprehensive understanding of the customer base. Within the marketing industry, this is referred to as customer relationship management (CRM), which will be discussed in more detail in subsequent chapters. When all customer transactions with the hotel or tourism company are available in a data warehouse, precise predictive models can be developed to anticipate future customer behaviour. These predictive models enable companies to create tailored products and messages for individual customers. Many travel and tourism organisations have developed predictive models for each product, which enable them to make informed inferences about future customer behaviour.

The primary focus in establishing data warehouses lies in data mining. Data mining involves extracting valuable information from the extensive pool of data stored within the warehouse. The Gartner Group defines data mining as the process of uncovering meaningful correlations, patterns, and trends by analysing vast amounts of data using pattern recognition technologies, statistical techniques, and mathematical algorithms. Data mining plays a crucial role in enabling managers and decision-makers to gather the necessary information to better understand customer behaviour and meet their needs effectively. Although setting up and maintaining data warehouses can be costly, it is essential to identify the significant data that should be collected and utilised in the decision-making process. While statistical sampling provides valuable insights, it still requires skilled managers who possess in-depth knowledge of their business to transform the data into actionable intelligence.

Direct Marketing

The concept of direct marketing has evolved over time, acquiring new meanings. Initially, it referred to a marketing approach where products or services were directly delivered from the producer to the consumer without intermediary involvement. This encompassed companies employing salespeople to engage in direct marketing. However, as technologies such as the telephone and other media gained prominence in directly promoting offers to customers, the Direct Marketing Association (DMA) redefined direct marketing as an interactive marketing system that utilises one or more advertising media to elicit a measurable response or transaction at any location. This definition highlights the emphasis on marketing efforts aimed at achieving a measurable response, typically in the form of a customer order. It is also known as direct-order marketing due to the nature of the transaction. Examples of direct marketing methods include email marketing and telemarketing (telephone-based marketing).

The introduction of customer database systems has enhanced the effectiveness and efficiency of direct marketing. By utilising the customer profiles stored in the database, companies can tailor their offers and communications to customers who are highly likely to be interested in and willing to purchase specific products or services. The utilisation of customer databases in the design and communication of offers significantly enhances the response rate of direct marketing campaigns.

Reasons for Growth of Direct Marketing

The advancement and widespread adoption of suitable technology have been crucial in the expansion of direct marketing. Particularly, the rise of computer-based technologies has significantly impacted various aspects of direct marketing. The integration of new computer technologies has enabled direct marketers to conduct more precise result analysis, target messages based on intricate psychographic and demographic factors, develop sophisticated databases of customers and prospects, and even enhance the creative execution of direct mail campaigns. The growth of technology, especially computer-based technologies, has played a significant role in elevating the effectiveness and efficiency of direct marketing practices.

The remarkable growth of direct marketing has been facilitated by the growing acceptance of telephone and internet channels for placing orders.

The availability of electronic ordering systems has not only accelerated order fulfillment but also eliminated the delays traditionally associated with mail orders. Presently, placing an order over the phone provides a level of "instant gratification" akin to purchasing an item directly from a store or obtaining a travel ticket from an agency. A similar trend is rapidly emerging for online purchases, although some consumers still express concerns about the perceived security risks associated with transmitting personal and financial information online.

Development of Integrated Direct Marketing

Many direct marketers typically depend on using a single advertising channel and a one-time attempt to reach and sell to prospects. A clear example of this approach is when a marketing campaign includes sending a single piece of mail that offers a weekend package at a hotel. This type of campaign is considered a single-vehicle, single-stage campaign, meaning that it relies on a single method of communication (in this case, mail) and doesn't involve any follow-up actions or additional stages. Alternatively, a single-vehicle, multiple-stage campaign involves sending a series of successive mailings to a prospect to encourage purchases. For instance, restaurants may send four notifications to a household to entice them to try their establishment. As mentioned earlier, Mauna Kea resort villas implemented a three-phase campaign. However, a more impactful strategy is to employ a campaign that involves multiple advertising channels and stages. This approach is referred to as integrated direct marketing (IDM). Let's examine the following sequence as an example:

Paid ad with a response channel	→	Direct mail mechanism	→	Outbound telemarketing	→	Face-to-face sales call

The paid advertisement aims at generating awareness of the product and stimulate inquiries. Once inquiries are received, the company follows up by sending direct mail. Within a period of 48–72 hours after the mail is received, the company contacts the prospects by phone to secure orders. Some prospects may make a purchase, while others may express interest in a face-to-face sales meeting. Even if a prospect is not yet prepared to make a purchase, ongoing communication is maintained. This approach, known as response compression, involves deploying multiple media channels within a tightly defined timeframe to maximise the impact and awareness of the message. The objective is to strategically employ specific media with precise timing to generate higher incremental sales while balancing the associated incremental costs. While a standalone direct-mail campaign may typically yield a 2 percent response rate, the use of integrated direct marketing can generate response rates of 12 percent or more.

Summary

In this chapter, we discussed the changing environment of the hospitality and tourism industry and the role of ICT development especially the internet and the world wide web in the changing environment. We then discussed internet marketing, database marketing, and direct marketing as results of the ICT's influence on marketing while enumerating ways by which each of the concepts can be applied for effective marketing management within the hospitality and tourism sector.

Activity

1. Define electronic marketing and explain any two reasons why is it relevant for the hospitality industry.
2. How can developments in ICT and the internet be leveraged for more effective marketing management in the tourism sector? Explain with practical examples.
3. Outline and discuss the process of developing effective websites in the hospitality industry.

References

Bala, M., & Verma, D. (2018). A critical review of digital marketing. *International Journal of Management, IT & Engineering*, *8*(10), 321–339.

Giao, H., Vuong, B., & Quan, T. (2020). The influence of website quality on consumer's e-loyalty through the mediating role of e-trust and e-satisfaction: An evidence from online shopping in Vietnam. *Uncertain Supply Chain Management*, *8*(2), 351–370.

Jensen, M. B. (2006). Characteristics of B2B adoption and planning of online marketing communications. *Journal of Targeting, Measurement and Analysis for Marketing, 14*, 357–368.

Kim, D. (2020). Internet and SMEs' internationalization: The role of platform and website. *Journal of International Management, 26*(1), 100690.

Kotler, P. (2003). *Marketing management* (11th ed.). New York, NY: Prentice-Hall.

Lustria, M. L. A. (2007). Can interactivity make a difference? Effects of interactivity on the comprehension of and attitudes toward online health content. *Journal of the American Society for Information Science and Technology, 58*(6), 766–776.

Middleton, V. T. C., Fyall, A., Morgan, M., & Ranchhod, A. (2009). *Marketing in travel and tourism* (4th ed.). Oxford: Elsevier.

Mohseni, S., Jayashree, S., Rezaei, S., Kasim, A., & Okumus, F. (2018). Attracting tourists to travel companies' websites: The structural relationship between website brand, personal value, shopping experience, perceived risk and purchase intention. *Current Issues in Tourism, 21*(6), 616–645.

Siegel, C. (2003). *Internet marketing: Foundations and applications*. Boston, MA: Houghton Mifflin.

Strauss, J., Frost, R., & Sinha, N. (2014). *E-marketing* (p. 496). Upper Saddle River, NJ: Pearson.

Chapter 24

Relationship Marketing

Chapter Outline

- Introduction
- Growth of Relationship Marketing
- Adoption of TQM
- Organisational Changes
- Increased Competition Intensity
- Building and Deepening Relationships
- Strengthening Relationships (Bonding)
- Customer Relationship Management (CRM)

Objectives

At the end of this chapter, readers should be able to:

- Discuss relationship marketing and its application in hospitality and tourism marketing.
- Identify stakeholders to build relationships with and explain the process of undertaking relationship marketing in relation to the stakeholders.
- Discuss the role of ICT in relationship marketing.

Introduction

Various developments in the micro and macro environments of industries have caused a shift from transactional marketing to relationship marketing.

DOI: 10.4324/9781032688497-24

Firms are therefore focusing on starting, developing, and maintaining profitable relationships with their stakeholders through relationship marketing. In this section, we examine the concept of relationship marketing and how it can be implemented within the hospitality and tourism industry context. We will also discuss CRM as ICT's contribution to improving the relationship marketing process.

The relationship a company has with its customers, channel members, and with competitors is an important source of competitive advantage since it is becoming obvious that competitive advantage could no longer be delivered on the basis of product or service characteristics alone (Buttle & Maklan, 2019; Casais et al., 2020). Most product or service offerings are similar and the quality is also becoming increasingly similar since the proliferation of providers in service or product categories has resulted in the demise or conformation to quality by poor product or service producers. Consequently, there is a paradigm shift from traditional marketing (transactional and mass marketing) to relationship marketing (Gronroos, 1996). As described by Gronroos (1996), relationship marketing involves recognising, establishing, sustaining, and, if needed, concluding profitable relationships with customers and other partners. The main goal is to ensure that the objectives of all parties involved are achieved and enhanced throughout the interactions (Table 24.1).

The primary attributes of relationship marketing can be succinctly summarised as follows:

- Concentrating on customer retention.
- Emphasising product benefits.
- Engaging in long-term interactions.
- Prioritising customer service.
- Encouraging active customer involvement.
- Maintaining frequent communication with customers.
- Implementing total quality management practices.

In terms of customers, relationship marketing emphasises the retention and long-term management of customer relationships to establish enduring and mutually beneficial connections. The core objectives are to retain customers and foster loyalty, making relationship marketing a crucial strategy for numerous organisations, including those in the hospitality and tourism sectors (Palmer, 2002).

Table 24.1 Differences between Transactions Marketing and Relationship Marketing

Transactions MarketingR	*Relationship Marketing*
Short-term orientation at sales as final result.	Long-term orientation at consumers where sales is just the beginning of the process.
"I" orientation.	"We" orientation.
Focus at selling projections.	Focus at keeping consumers and repeating of sales.
Stressing beliefs, persuasion to buy.	Stressing creation of mutual positive relations.
The need to reach selling goals, manipulation.	Reaching trust in services.
Stressing conflicts in realisation of transactions.	Partnership and cooperation in minimising lacks and providing long-term relationships with consumers, strategic partners, joint ventures, sellers.
Anonymous consumers are attracted through carefully planned events.	Individual consumer profile is known so the process could be developed on continual basis.

Source: Cooper C., Fletcher J., Gilbert D., Wanhill S., Shepherd, R., *Tourisms, Principles and Practice*, ed, Longman, New York, 2000, p.493.

Growth of Relationship Marketing

The resurgence of a relationship-oriented marketing approach signifies a revitalisation of direct connections between producers and consumers. Various environmental and organisational development factors have contributed to this renewed emphasis on direct relationships. Among these factors, at least five macro-environmental forces can be recognised:

- Swift progress in technology, particularly in the field of information technology;
- The implementation of Total Quality Management (TQM) initiatives by businesses;
- The expansion of the service economy;
- Organisational development procedures promoting the empowerment of individuals both within and outside the company;
- Heightened competitive pressure, leading to a focus on customer retention.

Rapid Technological Advancement

The technological revolution is reshaping the landscape of marketing institutions and their operations. The advent of advanced electronic and computerised communication systems has enabled consumers to engage directly with producers. Simultaneously, producers are now equipped with sophisticated databases that efficiently capture information from every consumer interaction, providing valuable insights into individual preferences. This capability empowers them to engage in personalised marketing practices at a cost-effective level.

Adoption of TQM

The technological revolution is reshaping the landscape of marketing institutions and their operations. The advent of advanced electronic and computerised communication systems has enabled consumers to engage directly with producers. Simultaneously, producers are now equipped with sophisticated databases that efficiently capture information from every consumer interaction, providing valuable insights into individual preferences. This capability empowers them to engage in personalised marketing practices at a cost-effective level.

Growth of the Service Economy

The rise of relationship marketing is influenced by the expansion of the service economy, particularly evident in advanced countries where the hospitality and tourism industry operates. A key factor contributing to this trend is the integrated nature of service production and delivery within the same institution. Service providers are directly involved in both producing and delivering their services. Take, for example, hotels, travel agencies, and Destination Management Organizations (DMOs) where the individuals responsible for providing the service are also the ones delivering it. As a result, stronger interactions and emotional connections develop between the service provider and the customer, emphasising the importance of fostering and improving these relationships.

Organisational Changes

Certain organisational transformations have played a crucial role in fostering the growth of relationship marketing. One of the most noteworthy changes involves redefining the roles of members within the organisation. Companies have introduced various alterations to their organisational processes to directly involve consumers in purchase decisions, acquisition, and even the design aspects of their products and services. In the past, these functions were typically managed by specialised departments without much input from the actual users of the products and services. However, this scenario has evolved, and now tourists, for instance, can actively participate in designing tour routes through platforms like websites or focus groups. As a result, the traditional separation between producers and users, often maintained by intermediaries, acting as gatekeepers, is potentially bridged in many cases. In these instances, direct interaction and collaborative relationships between producers and users naturally develop.

Increased Competition Intensity

In the era following industrialisation, the heightened level of competition has been compelling marketers to prioritise customer retention. Numerous studies have revealed that retaining existing customers is not only more cost-effective but also offers a potentially more sustainable competitive advantage compared to acquiring new customers. Marketers have come to recognise that it is more economical to retain existing customers than to invest in attracting new ones (Doaei et al., 2011; Aka et al., 2016; Palmer, 2002). From the supply side, developing closer relationships with a select group of suppliers is proving to be more beneficial than working with multiple vendors (Steinhoff et al., 2019). Furthermore, many marketers now prioritise cultivating lifelong customer relationships instead of focusing solely on one-time sales (Berry & Linoff, 2004) .

Types of Relationships

In the hospitality and tourism industry, particularly within the hospitality sector characterised by frequent interactions between consumers/tourists and selling personnel or companies, relationship marketing plays a critical role in the following segments:

- Between hospitality service retailers like hotels or airlines and intermediary marketing companies such as tour operators, incentive companies, and conglomerate travel agencies.

- Between hospitality service retailers and important customers such as large corporations and government agencies.
- Between food suppliers like McDonald's and organisations like universities, bus terminals, and large corporations.
- Among retailers operating within related tourist and hospitality service sectors, such as hotel chains, motels, and restaurants.
- Between hospitality service retailers and key suppliers.
- Between hotel organisations and their employees.
- Between hotel organisations and their marketing agencies, banks, law firms, etc.

Building and Deepening Relationships

When creating and implementing a relationship marketing strategy, the company's initial focus should be on enhancing the current relationships with customers. Enhancing the firm's connection with its existing customers can be achieved through four approaches:

(1) Enhancing the core service.
(2) Customising the relationship.
(3) Augmenting the services provided, and
(4) Implementing relationship-based pricing (Parasuraman et al., 1991).

The foundational elements described are interconnected and can be utilised together or concurrently. To begin, an excellent core service not only fulfils customers' needs but also establishes the business's reputation through its high quality, features, and long-term commitment. Additionally, it serves as a foundation for offering supplementary services as the customer relationship develops. When making relationship marketing decisions at this level, the objective is to fulfill the commitment of providing a flawless and impeccable core service to all clients. This is a crucial step in building and reinforcing customers' trust and confidence in the company's fundamental dependability and honesty.

The second aspect involves personalising the relationship with customers, necessitating the organisation to monitor, and comprehending the distinct attributes and needs of individual clients. Through database marketing, the company gains the capability to customise services based on each unique situation. In the hospitality sector, this technology enables comprehensive analysis of guest history, reservation databases, feedback from guest comment cards, preferences for additional services (like fine dining, spa treatments, golf, banqueting, corporate meeting spaces, etc.), information requests, and other sources of guest-related data.

One of the key considerations for a service firm involved in relationship customisation is its capability to establish "listening posts" within the organisation. These formal and informal communication channels are essential for recording guest preferences, complaints, and problem resolutions.

Database marketing plays a vital role in providing the necessary information for fundamental activities like customer profitability segmentation and targeted marketing. The underlying concept is that not all customers generate the same level of profit, and identifying the profitability of different market segments helps the firm decide where to allocate its limited resources to build loyalty among highly profitable customers. This type of segmentation also influences the level of service provided; typically, more profitable segments receive more personalised or customised services. For hospitality firms, the challenge lies in ensuring confidence in the data, including costs and potential revenues used to determine segment profitability, and making decisions about prioritising short-term or long-term profitability. For instance, a specific customer or customer segment might not be profitable at present, but its long-term potential might justify a short-term investment decision.

The phase of relationship customisation is a crucial moment for the company to assess the potential for repeat visits, signifying a profitable long-term relationship, and the effectiveness of service augmentation strategies (described below, the third foundation). The hotel endeavours to differentiate guests by calculating and comparing their customer lifetime value (LTV) and organising them into a segmented pyramid, with the most valuable clients at the top. LTV is a measure of how much a guest is worth to the hotel company and how long they will continue to contribute to its revenue.

The ultimate goal of all hotel frequency (loyalty) programmes, and ideally all hotels, is to maximise the LTV of their guests. However, from the customer's perspective, profitability segmentation, although logical, may not appear fair. Some customers may feel displeased receiving a lower level of service (e.g., automated instead of face-to-face service encounters) compared to what they previously received, especially if they perceive others are still receiving personalised and tailored services. This sentiment is particularly pronounced if they believe they are receiving less service than before or being asked to pay for services they once received free of charge.

The third pillar of relationship strategies is service augmentation, allowing the firm to set its services apart from competitors by offering unique ancillary services that hold value for the target market. Within the hotel industry, various forms of augmented services can be observed, such as new amenities and features, service quality guarantees, and preferred guest clubs. The benefits of these enhancements can be associated with guest comfort and increased confidence in the service quality, ultimately contributing to a positive self-perception.

Lastly, the foundation of relationship pricing involves the application of differential pricing, providing price incentives to repeat customers as a means of consolidating their business with a single service provider. Well-known examples of such relationships are frequent flyer and frequent guest programmes. While the price incentives can take different forms, like upgrades, free trips or room-nights, or special prices for spa access, the core objective remains consistent: to motivate customers to engage in repeat business by rewarding their loyalty.

Strengthening Relationships (Bonding)

After implementing the four fundamental principles for cultivating and enhancing relationships that ultimately result in customer retention, the hospitality company must determine the most suitable strategies to fortify connections with customers or partners. These specific strategies and tactics are employed to foster stronger relationships and forge a closer tie with clients, and they can be achieved by establishing various types of bonds.

(1) Financial bonds, (2) Social bonds, (3) Customisation bonds, and (4) Structural bonds (Zeithaml & Bitner, 2003).

The higher tiers of relationship strategies establish more robust connections with customers and pose greater challenges for competitors to replicate. Prominent hospitality companies frequently utilise multiple relationship strategies, potentially operating across all four levels concurrently.

Financial Bonds

A retention strategy grounded in financial bonds involves providing financial rewards to customers for repeated purchases or long-term loyalty. For instance, loyalty programmes like SAA Voyager and Shebamiles, operated by

South African and Ethiopian Airlines, respectively, offer benefits to frequent guests and passengers. Other incentives for repeat clients may include bundled services and cross-selling initiatives, either internally or in collaboration with other service providers, such as hotels and car rental companies, fine dining restaurants, and museums. Financial bonds can also be reinforced through stable pricing guarantees or by limiting price increases for loyal customers compared to new ones. However, solely relying on financial incentives does not typically lead to long-term, sustainable advantages for a firm, as it may not set the business apart from competitors in the long run, unless combined with another relationship strategy.

Social Bonds

This entails establishing social or interpersonal connections between the customer and the organisation. These bonds can vary from formal to informal relationships, where the hospitality firm endeavours to understand and cater to the individual needs and desires of each customer. Building social bonds requires a significant investment of time and effort. Crucial elements of social bonds are built on promises, trust, and commitment. The organisation and its employees must deliver on the promises made to customers, as failing to do so can strain or jeopardise the relationship. Trust, as defined by Moorman, is the confidence and willingness to rely on an exchange partner. It stems from the partner's competence and reliability, thus organisational personnel should demonstrate expertise and dependability. Commitment may take two forms: affective or calculative. Affective commitment arises from emotional attachment or liking, while calculative commitment is based on recognising the value of the relationship. Both types of commitment contribute to a mutually beneficial relationship, but affective commitment holds greater significance.

Customisation Bonds

These strategies are centred on establishing connections through personalised service. The underlying belief is that when customers receive tailored service that caters to their specific needs and circumstances, they will be more content and less likely to switch to competitors. Moreover, the time and effort invested by customers in educating a new service provider about their requirements make switching more challenging. Engaging in two-way communication with clients, supported by effective learning processes, becomes a sustainable means for anticipating and innovating services to suit individual customer needs. In-depth knowledge of individual customers and the creation of personalised solutions serve as the basis for other forms of customisation ties, such as mass-customisation and customer intimacy.

Structural Bonds

The structural type of bond, which a combination of all three types of bonds mentioned above is the most challenging type of bond between the customer and the firm. It is very difficult to replicate. Customers find it particularly difficult to disengage from these strategies because they also involve a structural aspect, often relying on shared systems or technology. Structural bonds are frequently observed in business-to-business interactions, where partners collaborate in processes and equipment sharing, joint investments, and the implementation of integrated information systems. For instance, the relationships between hotel companies and distributors illustrate noteworthy advancements in structural ties.

Customer Relationship Management (CRM)

Information technology (IT) plays a crucial role in enhancing company relationships with customers, facilitating customisation. This realisation has led companies to adopt customer-specific strategies instead of treating all customers uniformly. By focusing on the economically valuable customers, companies can employ a management approach that aims to enhance the retention of these valuable clientele (Soltani et al., 2018).

The increasing prevalence of information and communication technology (ICT) is enhancing the way business relationships are initiated, established, maintained, and concluded. It is playing a crucial role in improving the overall process of managing customer relationships, from the initial contact to ongoing maintenance and eventual conclusion. Companies are utilising ICT to gather and analyse data, predicting customer behavioural patterns, and equipping their staff with valuable information to interact with customers in a more effective and efficient manner.

Petzer et al. (2009) highlight that an organisation's success in the long term is closely linked to its capacity to establish positive relationships. Consequently, Customer Relationship Management (CRM) has emerged as a crucial aspect of hospitality marketing theory and practice. The growing significance of ICT in business operations, particularly data warehousing, has contributed to the viability of CRM as a means of conducting business effectively. The evolution of CRM can be traced back to relationship marketing, which places a higher emphasis on enhancing customer retention through efficient management of customer relationships.

To effectively employ CRM systems, companies within the hospitality and tourism industry need to continuously collect and analyse customer data. This crucial data could be collected from customers at vantage contact points with customer such as during reservation, check-in, checkout, and potentially throughout their stay. Subsequently, this data can be analysed and utilised to enhance the firm's services. For CRM systems to be efficient, they should encompass data on marketing aspects such as customer satisfaction or dissatisfaction, behavioural loyalty encompassing purchase frequency and volume over time, and customer feedback and complaints.

In summary, CRM is a multidimensional approach that fosters ongoing communication with customers across various touchpoints and interactions. It involves personalised interactions with high-value customers, aiming to optimise revenue and enhance customer value by comprehending and fulfilling individual customer requirements (Kumar & Reinartz, 2018).

Activity

1. Discuss how you will implement relationship marketing in a chosen hospitality or tourism firm and the role of ICT in improving the relationship marketing strategy.
2. Explain the importance of relationship marketing in the tourism industry and its effects on its profitability.
3. Provide three differences between transactional and relationship marketing.
4. Discuss the four main strategies through which managers in the hospitality industry could build, bond, and strengthen the relationship with their clients.

References

Aka, D., Kehinde, O., & Ogunnaike, O. (2016). Relationship marketing and customer satisfaction: A conceptual perspective. *Binus Business Review, 7*(2), 185–190.

Berry, M. J., & Linoff, G. S. (2004). *Data mining techniques: For marketing, sales, and customer relationship management.* John Wiley & Sons.

Buttle, F., & Maklan, S. (2019). *Customer relationship management: Concepts and technologies.* Routledge.

Casais, B., Fernandes, J., & Sarmento, M. (2020). Tourism innovation through relationship marketing and value co-creation: A study on peer-to-peer online platforms for sharing accommodation. *Journal of Hospitality and Tourism Management, 42*, 51–57.

Doaei, H., Rezaei, A., & Khajei, R. (2011). The impact of relationship marketing tactics on customer loyalty: The mediation role of relationship quality. *International Journal of Business Administration, 2*(3), 83.

Gronroos, C. (1996). Relationship marketing: Strategic and tactical implications. *Management Decision, 34*(3), 5–14.

Kumar, V., & Reinartz, W. (2018). *Customer relationship management.* Springer-Verlag GmbH, part of Springer Nature 2006, 2012, 2018.

Palmer, A. (2002). The evolution of an idea: An environmental explanation of relationship marketing. *Journal of Relationship Marketing, 1*(1), 79–94.

Parasuraman, A., Berry, L. L., & Zeithaml, V. A. (1991). Perceived service quality as a customer-based performance measure: An empirical examination of organizational barriers using an extended service quality model. *Human Resource Management, 30*(3), 335–364.

Petzer, J. P., Castagnoli, Jr. N., Schwarzschild, M. A., Chen, J. F., & Van der Schyf, C. J. (2009). Dual-target–directed drugs that block monoamine oxidase B and adenosine A2A receptors for Parkinson's disease. *Neurotherapeutics, 6*(1), 141–151.

Soltani, Z., Zareie, B., Milani, F. S., & Navimipour, N. J. (2018). The impact of the customer relationship management on the organization performance. *The Journal of High Technology Management Research, 29*(2), 237–246.

Steinhoff, L., Arli, D., Weaven, S., & Kozlenkova, I. V. (2019). Online relationship marketing. *Journal of the Academy of marketing science, 47*, 369–393.

Zeithaml, V. A., & Bitner, M. J. (2003). *Services marketing: Integrating customer focus across the firm* (3rd ed.). Tata McGraw-Hill.

Chapter 25

Destination Marketing

Chapter Outline

- Introduction
- Tourist Destination
- Management of Tourist destinations
- Destination's Appeal
- Destination Image
- Affordable Packages
- Infrastructure and Security
- Sustainable Tourism

Objectives

At the end of this chapter, the student should be able to:

- Discuss the influences of tourism on national development.
- Discuss the effective management of a tourism destination.
- Explain the concept of sustainable tourism.

Introduction

Despite the COVID-19 pandemic, tourism continues to exert significant influence on the global economy, offering numerous opportunities for countries

DOI: 10.4324/9781032688497-25

to engage in this socioeconomic phenomenon. The rise of tourism has brought about substantial changes in destination management and marketing (Fyall & Garrod, 2020). As destinations strive to remain competitive, they face the challenge of enhancing their appeal. This chapter delves into the growth of the tourism sector and its impact on national development and destination marketing. Additionally, we explore effective destination management and the role of sustainable practices in preserving destinations through sustainable tourism. Over the past three decades, tourism has emerged as a dominant player in the global economy, providing increasing prospects for both developed and developing countries to participate in this socioeconomic activity.

The competition to attract tourists has intensified as destinations worldwide undergo rapid tourism developments, aiming to capitalise on the economic advantages offered by this thriving industry. Tourism demand has experienced a remarkable surge, with international arrivals rising from 565 million in 1995 to over 1.5 billion in 2019, and is expected to bounce back sharply after dropping in 2020 due to COVID-19 (Statista, 2023).

The tourism industry's growth has driven significant transformations in the management and marketing of destinations. For many destinations, the challenge lies in achieving or maintaining competitiveness. According to Fourie et al. (2022), a truly competitive tourism destination is one that can increase tourism expenditure, attract visitors while providing them with satisfying, memorable experiences, and do so profitably while benefiting local residents and preserving the destination's natural resources for future generations. Destination competitiveness refers to the destination's ability to attract and satisfy tourists by offering products and services perceived to be superior to those of other competing destinations. Marketing plays a crucial role in creating competitive destinations. With this context in mind, destination marketing assumes a critical role in preventing the destination life cycle from entering a stage of saturation and decline and enables the destination to adapt to the changing marketplace, seize opportunities, and maintain its vibrancy (Sorokina et al., 2022).

Destination marketing involves promoting a destination with the ultimate aim of increasing visitor numbers. According to Ketter (2018), the scope of destination marketing encompasses a management process in which national tourist organisations and/or tourist enterprises identify their selected tourists, both current and potential, engage in communication to understand and influence their preferences, needs, motivations, likes, and dislikes on local, regional, national, and international levels. The objective is to formulate and adapt tourist products accordingly, with the goal of achieving optimal tourist satisfaction and fulfilling their objectives.

The travel and tourism sector is one of the most dynamic and successful industries in the global economy. Whether discussing positive or negative impacts, tourism's influence is felt across a significant portion of global society in various aspects, including its effects on the national economy, culture, social dynamics, and the environment. The tourism industry has a significant impact on the world's economy, evident in its substantial contribution of USD2.6 trillion to global GDP and its role in providing approximately 18 percent of the world's employment opportunities (WTTC, 2023).

The tourism industry stands out as the fastest-growing sector globally. One of the critical concerns in this industry is assessing tourist satisfaction, especially concerning the destination they visit. Recognising the vital role of consumer satisfaction, various companies in the tourism sector conduct regular consumer satisfaction surveys to gauge the level of satisfaction among their customers. These surveys are sometimes carried out independently or with the assistance of marketing agencies. The significance of tourism for economies lies in the diverse range of activities it encompasses, including travel agencies, accommodations, sightseeing, shopping, and entertainment (Fourie et al., 2022). Due to the importance of tourist destination attributes, numerous studies have been conducted to examine destination image.

In the present day, destinations are confronted with more intense competition than ever before. Major tourist destinations worldwide are vying for a share of the tourism market. In this highly competitive landscape, the competitiveness of a tourism destination relies on its ability to develop effective marketing and management strategies. Hence, Baloglu et al. (2004) argue that adopting "one-for-all" models, which assume that the overall tourist population is homogenous, would not be practical or useful for a destination in such a fiercely competitive environment.

Tourist Destination

Tourist destinations can be described in different ways, whether by their physical, political, or market-defined boundaries. They are intricate systems comprising interconnected elements like attractions, accommodations, transportation, and various supporting services. The significance of destinations in driving tourism demand cannot be overstated. However, to tap into this potential, all stakeholders within the destination must collaborate to stay competitive in the constantly evolving tourism market. A destination's offering encompasses a diverse mix of services, facilities, and infrastructure, including shopping, dining, transportation, accommodations, attractions, and entertainment. The presence of reliable infrastructure and political stability are vital factors that influence a destination's ability to attract tourists. Some destinations are classified as macro destinations, like the United States, which encompass numerous micro destinations within them. These micro destinations can range from regions, states, cities, towns, and even specific attractions within a town. Together, they contribute to the overall appeal and attractiveness of the larger macro destination.

Management of Tourist Destinations

In practical terms, destination marketing is a strategic process that leverages competitive advantages, targets specific markets, employs a mix of marketing techniques, and appeals to both current and potential visitors. This strategy involves aligning the destination's resources with market opportunities. Each destination should strive to set itself apart by showcasing its unique tangible and intangible offerings, making it enticing for visitors to choose it as their preferred destination.

In today's highly competitive global tourism market, merely attracting visitors for a one-time visit is not enough. Destinations must aim to foster loyalty among their visitors, encouraging repeat visits and positive word-of-mouth recommendations. Given the dynamic global landscape, destinations must take a proactive, resilient, and strategic approach to marketing.

Tourism managers face the challenge of understanding why travellers choose a specific destination amidst countless options and how to ensure these individuals become repeat visitors and advocates for the destination. Some destinations have excelled in this aspect, like France and the United Kingdom in the North and Mexico in the South, which have achieved significant success in attracting and retaining tourists.

Destination Appeal

Destinations can be described as specific places or areas that offer a range of facilities and services to satisfy the needs of tourists (Bigne et al., 2019). According to the conceptual model of destination competitiveness by Crouch

(2007), the core resources and attractors of a destination play a crucial role in influencing tourists to choose one destination over another. These core resources and attractors act as primary motivations or pull factors for inbound tourism. Crouch (2007) identify several core pull factors at the destination, including physiography (natural landscapes and features), culture and history, a diverse mix of activities, special events, entertainment options, and the availability of necessary infrastructure and amenities. The success of these destinations is attributed to a combination of factors, such as effective imaging and branding to create a distinct identity, the development of affordable travel packages to attract a wider audience, and the continuous maintenance and enhancement of the destination's offerings to ensure visitor satisfaction and loyalty.

Destination Image

In the highly competitive tourism market, the concept of destination image has emerged as a critical aspect of effective tourism management and destination marketing (Afshardoost & Eshaghi, 2020). Destination image, as defined by Rasoolimanesh et al. (2021), refers to the visual or mental impression that the general public has of a place or product. The perceived brand image plays a significant role in shaping consumer behaviour. It allows potential customers to recognise a product, assess its quality, reduce purchase risks, and anticipate certain experiences and satisfaction through product differentiation (Afshardoost & Eshaghi, 2020). Consumers often use brand image as an external cue when making purchasing decisions (Rasoolimanesh et al., 2021; Pan et al., 2021). It is essential to distinguish between two types of destination images: organic and induced. Organic destination image arises naturally from how potential tourists perceive a destination without any deliberate marketing efforts by destination marketers. It is influenced by the destination's inherent attractions, climate, history, and culture, and media portrayals of the destination. In contrast, induced destination image is intentionally created by destination management organisations to attract visitors. Establishing an induced destination image requires substantial marketing efforts to showcase the destination's unique selling points. Tourism organisations employ various marketing communication strategies to craft and promote an induced destination image.

A destination with a positive image has the advantage of reducing the perceived risks that potential customers might encounter when making their travel decisions (Afshardoost & Eshaghi, 2020). However, creating a brand image typically involves promotional efforts, advertisements, and the overall experiences associated with the destination (Foroudi et al., 2018; Pan et al., 2021). Rasoolimanesh et al. (2021) pointed out that potential tourists, who often have limited knowledge about the destination and no prior visit experience, are more likely to choose destinations with strong, positive, and recognisable images.

Since the destination image significantly influences tourists' perceptions, it becomes crucial to understand the factors that contribute to the formation of this image in the minds of travellers. Implementing effective strategies that consistently communicate and foster the creation of the right image in the minds of tourists is vital for destinations to attract and retain visitors successfully.

Affordable Packages

An essential aspect of destination management involves the creation of well-designed and effectively communicated tourism packages. These packages encompass a carefully coordinated system that includes various tourist sites and events within a specific destination, along with accommodations, transportation, food, entertainment, and a diverse range of activities. To develop these enticing packages, destination managers must gather relevant data on tourists' travel patterns, lodging preferences, transportation choices, and other pertinent tourist preferences. By analysing and interpreting this data, destinations can tailor attractive packages that cater to the interests and needs of both current and potential tourists. These well-crafted packages play a pivotal role in enhancing the overall tourism experience, increasing visitor satisfaction, and ultimately contributing to the destination's success in the competitive tourism market.

Infrastructure and Security

Destinations that neglect to maintain adequate infrastructure or develop inappropriate facilities face significant risks. For instance, some of East Africa's renowned game parks are experiencing degradation due to tourists being ferried around in four-wheel-drive vehicles, leading to dust bowls. Various factors can diminish a destination's attractiveness, such as violence, political instability, natural disasters, adverse environmental conditions, and overcrowding. A prominent example is Greece's national treasure, the once-white marble Parthenon in Athens, now tainted by pollution as a symbol of environmental neglect. Similarly, Thailand's beautiful beach resorts and temples have suffered considerable damage from pollution and poor sanitation. In the case of India, the government's plans for a "Visit India Year" were hindered not only by sectarian and caste violence but also by plane crashes. Consequently, Western countries, including the United States and Japan, deemed India an unsafe destination for their citizens. These examples illustrate the vital importance of maintaining and preserving a destination's infrastructure and environment to safeguard its attractiveness and ensure a positive experience for visitors.

Afshardoost and Eshaghi (2020) emphasise the vital role of destination marketing in cultivating and sustaining the popularity of a specific location. They express concern that, regrettably, many tourism planners tend to concentrate solely on developing new attractions and amenities without giving due attention to the conservation and preservation of the unique attributes that originally drew travellers to the destination. In doing so, there is a risk of losing the essence and allure that initially made the location attractive to tourists. Therefore, an effective destination marketing strategy should not only focus on attracting visitors but also on preserving the authentic appeal and distinctive features that contribute to the destination's appeal in the first place.

Sustainable Tourism

Sustainable tourism has emerged as a critical and integral aspect of the tourism industry. In recent times, it has become the central topic of discussion in the context of ethical travel (Higgins-Desbiolles, 2018). While many perspectives on sustainable tourism focus on addressing environmental concerns, only a limited number of interpretations take into account the three dimensions of sustainability, namely economic, social, and environmental aspects (Streimikiene, et al., 2021). Sustainable tourism endeavours to strike a balance between promoting economic growth, fostering social inclusivity and cultural preservation, and mitigating environmental impacts. As tourism continues to evolve, a comprehensive approach to sustainability that encompasses all these dimensions is becoming increasingly vital for the long-term well-being of destinations and the satisfaction of travellers.

Streimikiene et al. (2021) presents a comprehensive and encompassing definition of sustainable tourism as tourism activities and associated infrastructures that preserves the natural capacity of the ecosystem to regenerate and sustain future productivity of natural resources. Furthermore, sustainable tourism acknowledges the valuable contribution made by people, communities, customs, and lifestyles in shaping the tourism experience. It emphasises the importance of ensuring that these local stakeholders have fair and equitable access to the economic benefits generated by tourism activities within the host areas. In essence, sustainable tourism seeks to strike a balance between environmental preservation, social inclusivity, and economic prosperity, both for the present and future generations (Rai, 2012).

The World Tourism Organization defines sustainable tourism as a form of tourism that not only caters to the needs and desires of current tourists and host regions but also actively safeguards and enriches opportunities for future generations. This type of tourism places significant emphasis on the role and responsibility of individuals involved, including both the residents of the destination and the tourists themselves, in making ethical and sustainable choices. The ultimate aim of sustainable tourism is to ensure that tourism activities contribute positively to environmental conservation, social inclusivity, and economic development, while also preserving the cultural heritage and natural resources of the destination for the long-term benefit of all stakeholders.

The core objective of sustainable tourism is to preserve the sociocultural authenticity of host communities, while also ensuring equitable socioeconomic benefits for all involved, promoting long-term economic viability, and optimising the use of environmental resources through conservation efforts. Sustainable tourism strives to harmoniously integrate social responsibility, economic efficiency, and ecological sensitivity at every stage of its development. The host country must safeguard its natural, historical, and cultural resources, vital ecological processes, biological diversity, and species protection to ensure the sustainability of tourism (Higgins-Desbiolles, 2018; Streimikiene et al., 2021).

In essence, sustainable tourism seeks to contribute positively to both the economy and society, while ensuring the responsible and sustainable use of all resources and the environment. It also aims to create employment opportunities and provide essential social services to local communities (Sustainable Development of Tourism Conceptual Definition, WTTC, 2023). The approach to sustainable tourism also recognises the importance of involving all stakeholders to foster consensus-building and ensure that those impacted by changes can voice their opinions about these transformations.

Developing sustainable tourism means acknowledging the interconnectedness of people's well-being and the well-being of the environment. Poverty is identified as a significant driver of environmental degradation; therefore, effective sustainable tourism projects must actively work towards poverty eradication in host destinations. By addressing these interlinked challenges, sustainable tourism aims to create a more inclusive, equitable, and environmentally conscious form of tourism that benefits both present and future generations.

To ensure a destination's sustainability in tourism, it is essential to monitor and manage the environmental, economic, and social impacts, striving to maintain them at acceptable levels for both visitors and business interests. Getz (1983) highlighted six capacity criteria, which include physical (tangible resources), economic, ecological, social/cultural, political/administrative, and perceptions of visitors, to identify potential thresholds that must be considered.

Among the various forms of sustainable tourism, ecotourism plays a significant role by safeguarding natural, environmental, and sociocultural resources. It aims to integrate continuous and sustainable practices within its operations to minimise negative impacts. The International Ecotourism Society (TIES) defines ecotourism as a type of tourism that focuses on preserving the environment, enhancing the well-being of local communities, and promoting responsible travel to natural areas.

Summary

In this section, we discussed the growth in tourism and its influence on the development of a nation. We then discussed the effective management of destinations by developing and maintaining a destinations appeal. Finally, we discussed the emerging concept of sustainable tourism.

Activity

1. Explain destination marketing and how managers of tourist destinations could make the destinations attractive and induce visitation.
2. Discuss the influences of tourism on national development.
3. Discuss the concept of sustainable tourism and its role in destination marketing.

References

Afshardoost, M., & Eshaghi, M. S. (2020). Destination image and tourist behavioural intentions: A meta-analysis. *Tourism Management, 81*, 104154.

Baloglu, S., Pekcan, A., Chen, S. L., & Santos, J. (2004). The relationship between destination performance, overall satisfaction, and behavioral intention for distinct segments. *Journal of Quality Assurance in Hospitality & Tourism, 4*(3–4), 149–165.

Bigne, E., Ruiz, C., & Curras-Perez, R. (2019). Destination appeal through digitalized comments. *Journal of Business Research, 101*, 447–453.

Crouch, G. I. (2007). Modelling destination competitiveness. *A survey and analysis of the impact of competitiveness attributes.* CRC for Sustainable Tourism Pty Ltd: Queensland, Australia.

Foroudi, P., Akarsu, T. N., Ageeva, E., Foroudi, M. M., Dennis, C., & Melewar, T. C. (2018). Promising the dream: Changing destination image of London through the effect of website place. *Journal of Business Research, 83*, 97–110.

Fourie, A., van Heerden, C., & Du Plessis, E. (2022). Improving destination competitiveness in South Africa: A DEA approach. *Tourism Economics, 28*(4), 1080–1100.

Fyall, A., & Garrod, B. (2020). Destination management: A perspective article. *Tourism Review, 75*(1), 165–169.

Getz, D. (1983). Capacity to absorb tourism: Concepts and implications for strategic planning. *Annals of tourism Research, 10*(2), 239–263.

Higgins-Desbiolles, F. (2018). Sustainable tourism: Sustaining tourism or something more? *Tourism Management Perspectives, 25*, 157–160.

Ketter, E. (2018). It's all about you: Destination marketing campaigns in the experience economy era. *Tourism Review, 73*(3), 331–343.

Pan, X., Rasouli, S., & Timmermans, H. (2021). Investigating tourist destination choice: Effect of destination image from social network members. *Tourism Management, 83*, 104217.

Rai, A. K. (2012). *Customer relationship management: Concepts and cases.* PHI Learning Pvt. Ltd.

Rasoolimanesh, S. M., Seyfi, S., Rastegar, R., & Hall, C. M. (2021). Destination image during the COVID-19 pandemic and future travel behavior: The moderating role of past experience. *Journal of Destination Marketing & Management, 21*, 100620.

Sorokina, E., Wang, Y., Fyall, A., Lugosi, P., Torres, E., & Jung, T. (2022). Constructing a smart destination framework: A destination marketing organization perspective. *Journal of Destination Marketing & Management, 23*, 100688.

Statista. (2023). Number of international tourist arrivals worldwide 1995–2022. Retrieved from: https://www.statista.com/statistics/209334/total-number-of-international-tourist-arrivals/

Streimikiene, D., Svagzdiene, B., Jasinskas, E., & Simanavicius, A. (2021). Sustainable tourism development and competitiveness: The systematic literature review. *Sustainable Development, 29*(1), 259–271.

WTTC. (2023). Travel & Tourism sector nears pre-pandemic recovery despite lengthy restrictions. Retrieved from https://wttc.org/news-article/japan-eir-2023#:~:text=WTTC%20forecasts%20the%20region's%20GDP,16%25%20below%20the%202019%20highpoint

Chapter 26

Segmenting and Monitoring the Tourist Market

Chapter Outline

- Introduction
- Market Segmentation
- Information Sources for Segmentation
- Identifying Segments for Targeting
- Criteria for Selecting Segment for Targeting
- Classification of Tourists
- Visiting Friends and Relatives (VFR)
- Monitoring the Tourist Markets

Objectives

At the end of this chapter, the student should be able to:

- Discuss the concept, reasons, and benefits of segmentation.
- Understand the process for selecting a segment for targeting.

Introduction

Increasing competition in the hospitality and tourism industry has increased the need for firms to focus their attention on specific segments of the

DOI: 10.4324/9781032688497-26

available market whose needs they can best satisfy for value. The benefit of market segmentation, therefore, lies in a tourist destination being able to specialise in the needs of a particular group and become the best in catering to that group. In doing so, the destination gains a competitive advantage. In this chapter, we will discuss the related concepts of segmentation and targeting.

Market Segmentation

New destinations appear every year. To have any chance of standing out in a busy marketplace, destinations must be single-minded focusing on those people who they are most likely to be able to attract and who are most likely to purchase a destination's tourism goods and services. Market segmentation is the starting point for devising a marketing strategy and is a process that categorises people into groups where they share certain definable characteristics.

There are several reasons to segment a market and these include:

- Only a small percentage of the world's population takes an international trip in any one year.
- Demand for tourism goods and services is not equally distributed throughout a population.
- Minorities within a population are often likely to consume a disproportionately high volume of tourism products.
- It is necessary to develop the product according to the customer's needs, wants, and desires.
- Segmentation helps inform which marketing channels and media are the right ones.

The benefit of market segmentation lies in a tourist destination being able to specialise in the needs of a particular group and become the best in catering to this group. In doing so, the destination gains a competitive advantage because:

(1) Competition can be reduced from the global market to tourism destinations specialising on the same segment (e.g., all ecotourism destinations).

(2) Efforts can be focused on improving the product in a specific way rather than trying to provide all things to all people at high cost (e.g., a family destination is unlikely to need extensive nightlife options).
(3) Marketing efforts can be focused on developing the most effective message for the segment targeted (e.g., a sun and fun message for young tourists travelling with friends) and by communicating the message through the most effective communication channel for the segment (e.g., in national geographic or other nature magazines for ecotourists).
(4) Tourists experiencing a vacation at a destination that suits their special needs would more likely be satisfied with their stay and, consequently, revisit and advertise the destination among like-minded friends.

The key to effective target marketing is to:

- Identify groups of people (segments) who are, or will be, in the market for an inter-national trip.
- Decide whether these are the people who, if they visited, would help fulfil the destination's tourism objectives.
- Establish whether the destination has the appropriate products and services to meet their needs.
- Establish their motivations/triggers, buying habits, etc.
- Persuade them to visit using appropriate and targeted marketing messages and channels.
- Evaluate and review the impact of the marketing on the segments that have been targeted (Weaver and Lawton, 2007).

Information Sources for Segmentation

The foundation for good segmentation is based on the collection of inbound tourism statistical data. Tourism visitor surveys would appear to provide the most fruitful source of identifying current tourism segments that visit the host destination. Data collection must however be strategic, taking into consideration the information needed to make the right decisions in relation to segmentation. The most common factors that can be included in statistical collection and analysis for segmentation include:

- Activities undertaken by tourists (cycling, hiking, golf, etc.).
- Booking mechanism (internet, travel agent).

- Cohort (with whom they travel).
- Demographic data (age, occupation, income, education, family, etc.).
- Group/party size.
- Length of stay.
- Motivations (recreation, culture, adventure, sun-seeking, etc.).
- Origin country/place of residence.
- Other countries considered for a visit (i.e., information about competitors).
- Purpose of visit (business, leisure, education, special events, etc.).
- Repeat or first-time visitor.
- Sources of information when planning to visit a country.
- Spending (by category: accommodation, travel, etc.).
- Transport into country (air, road, sea, rail).
- Transport used while in country.
- Travel dates (by month to assess seasonal variations).
- Trip type (independent or package).
- Type of accommodation used. (WTO, 2007a)

Identifying Segments for Targeting

Once a firm is in possession of the basic statistical and qualitative data required to disaggregate its source market(s), it is in a position to begin a process of segmentation, which means identifying those groups of customers within a market(s) that share a range of common characteristics and selecting those that it wishes to target.

Having identified a range of segments, firms are then faced with the challenge of deciding which of these segments offers the best prospects for contributing to the tourism objectives of their destinations. To do this, some firms can use a range of factors to compare the relative attractiveness of segments. Some of the factors most commonly used as a basis for comparing segments are:

- Accommodation capacity of the destination.
- Awareness and image of the destination.
- Income and education levels of tourist segments.
- Market share.
- Media usage and availability of promotional opportunities.
- Value/spending of the tourist segment (current and forecast).

However, the expected outcome from market segmentation is a competitive advantage. Consequently, the actual segmentation task aims to group tourists in a way that is of the most managerial value. For a segment to be managerially useful several requirements should be fulfilled.

Criteria for Selecting Segments for Targeting

1. The segment should be distinct, meaning that members of one segment should be as similar as possible to each other and as different as possible from other segments.
2. The segment should match the strengths of the tourism destination.
3. The segment should be identifiable. While female travellers can be identified very easily, identification of other visitors who are motivated by rest and relaxation may not be identified easily.
4. The segment should be reachable in order to enable destination management to communicate effectively. For instance, surf tourists are likely to read surf magazines which could be used to advertise the destination.
5. A segment should be suitable in size. This does not necessarily imply that a bigger segment is better. A tourism destination may choose to target a small niche segment that represents a large enough market for the particular destination and has the advantage of having very distinct requirements.

The above criteria for the usefulness of segments have to be considered when one or more of many possible segments are chosen for active targeting. Market segments can be derived in many different ways. All segmentation approaches can be classified as being either *a priori (common sense)* segmentation approaches (Dolničar, 2004; Mazanec, 2000) or *a posteriori (post hoc, data-driven)* segmentation approaches (Dolničar, 2004; Mazanec, 2000; Myers and Tauber, 1977). The names are indicative of the nature of these two approaches. In the first case, destination management is aware of the segmentation criterion that will produce a potentially useful grouping (common sense) in advance before the analysis is undertaken (a priori). In the second case, destination management relies on the analysis of the data (data-driven) to gain insight into the market structure and decides after the analysis (a posteriori, post hoc) which segmentation base or grouping is the most suitable one.

Once target markets are selected, firms move into developing marketing strategies for these markets. Some run campaigns/activities aimed at similar segments in different markets.

Classification of Tourists

There are various ways in which tourists are classified as a description of the segment into which they belong (Dilmonov, 2020). *Cohen's classification of tourists* described four types based on the institutionalisation of the tourist and the nature of the impact on the host community. There are four main categories that exist under *Cohen's Classification of tourists* as follows:

The Organised Mass Tourist

These are the least adventurous tourists, who on buying their package holiday remain encapsulated in an "environmental bubble," and usually view their destinations through the windows of the tour bus, divorced from the host community as they remain in the hotel complex. They adhere to an itinerary fixed by the tour operators, and even their trips out of the complex are organised tours. They make few decisions about their holiday.

The Individual Mass Tourist

They are similar to the organised mass tourist in that they utilise the facilities made available by the tour operator, but they have some control over their own itinerary. They may use the hotel as a base and hire a car for their own trips.

The Explorer

The explorers arrange their trip alone and attempt to get off the beaten track. Yet they will still have recourse to comfortable tourist accommodation. However, much of their travel will be prompted by a motivation to associate with the people, and they will often speak the language of the host community. Nonetheless, the explorers retain many of the basic routines of their lifestyle.

The Drifter

These people, the backpacker group, will seldom if ever be found in a traditional hotel. They may stay at youth hostels with friends or camp out. They tend to mix with lower-socioeconomic native groups and are commonly found riding third-class rail or bus. Most tend to be young.

Other accepted forms of classification of tourists who are observed around the world include:

Visiting Friends and Relatives (VFR)

This segment comprises of tourists whose primary reason for travelling to the destination is to visit friends and relatives. Segmentation has been radically re-assessed and is now seen as a major source of international tourism revenue and is set to increase due to the radical population shifts affecting the modern world. The various classes of VFRs are showcased next:

- Grief travel

Airlines offer special fares for family and friends attending funeral services. This segment will increase in importance as society ages.

- Business travellers

This often encompasses any form of business including conventions, trade shows, job seeking, and many other reasons.

- Leisure travel

This is a very wide and all-encompassing classification of tourists who visit a destination mainly for pleasure and relaxation; it may be of limited use without further segmentation.

- Business and pleasure travellers

Many conventions and business travellers plan to incorporate a period of relaxation prior to or after their business.

- Tag-along visitors

Members of the family are common "tag-along" visitors. The presence of tag-along children has created a sub-industry of child care and entertainment.

Monitoring Tourist Markets

Tourist markets are dynamic, and a marketing information system is part of any well-run tourist organisation. Destinations need to closely monitor the relative popularity of their various attractions by determining the number and type of tourists attracted to each. The popularity of a destination can suddenly or gradually change.

Marketing information systems help to identify and predict environmental trends that are responsible for these changes. Information should be collected on the changes in the wants of existing markets, emerging markets, and potential target markets. For example, in the mid-1990s the Japanese yen appreciated relative to Western currencies. Australia, a popular destination for the Japanese, now became even more of a value because of the increased purchasing power of Japanese tourist. The state of Queensland increased its marketing efforts towards the Japanese to take advantage of the favourable currency exchange.

Activity

1. Explain tourist segmentation and provide any three benefits of segmenting tourist destination.
2. Discuss the role of segmentation and targeting in the overall marketing strategy of a hospitality and tourism firm.
3. Explain target marketing and the process of achieving effective target marketing in the tourism sector.
4. The foundation for good segmentation is based on the collection of inbound tourism statistical data. Outline six sources of information for effective segmentation in tourist destination marketing.

References

Dilmonov, K. B. (2020). Classification and types of tourism. *International Scientific Review*, (LXX).

Dolničar, S. (2004). Beyond "commonsense segmentation": A systematics of segmentation approaches in tourism. *Journal of Travel Research, 42*(3), 244–250.
Mazanec, J. A. (2000). Mastering unobserved heterogeneity in tourist behavior research. *Tourism Analysis, 5*(2–3), 171–176.
Myers, J. H., & Tauber, E. (1977). *Market structure analysis.* Chicago: American Marketing Association
Weaver, D. B., & Lawton, L. J. (2007). 'Just because it's gone doesn't mean it isn't there anymore': Planning for attraction residuality. *Tourism Management, 28*(1), 108–117.
World Tourism Organization. (2007). *Handbook on tourism market segmentation, maximizing marketing effectiveness.* Madrid: World Tourism Organization and European Travel Commission.

Chapter 27

Tourism Promotion Strategies and Investments

Chapter Outline

- Introduction
- Attractions
- Events Marketing
- A Decision Framework: Event Planning

Objectives

At the end of this chapter, readers should be able to:

- Identify various systems and infrastructure needed to enhance the attractiveness of destinations.
- Discuss the development of attractions as investments in tourism.
- Discuss events and employ a decision framework for planning events.

Introduction

The dynamic and increasingly competitive environment of the hospitality and tourism industry has deepened the need for companies and governments to develop strategies and invest in infrastructure as a way of

DOI: 10.4324/9781032688497-27

increasing the attractiveness of destinations (Cohen, 1979). These strategies and investments are usually in the form of natural or man-made attractions and other supporting systems and infrastructure. In this chapter, we will discuss the development of these strategies and investments with a focus on attractions.

To attract tourists, destinations must respond to the travel basics of cost, convenience, and timeliness. Like other consumers, tourists weigh costs against the benefits of specific destinations and investment of time, effort, and resources against a reasonable return in education, experience, fun relaxation, and memories. Convenience takes on various meanings in travel decisions: time involved in travel from the airport to lodging, language barriers, cleanliness and sanitary concerns, access to interests (beaches, attractions, amenities), and special needs (elderly, disabled, children, dietary, medical care, fax, and communication, auto rental). Timeliness embraces factors that introduce risk to travel such as civil disturbances, political instability, currency fluctuations, safety, and sanitary conditions.

Places are therefore increasingly developing attractions and events as a vital component in attracting tourists (Mair and Weber, 2019). Small or rural places typically initiate an event such as a festival to establish their identity. Urban newspapers and suburban weeklies often publish a list of events, festivals, and celebrations occurring within a day's driving distance. State and local tourism offices do the same, making sure that travel agents, restaurants, hotels, airports, and train and bus stations have event-based calendars for posting. Tourism investment ranges from relatively low-cost market entry for festivals or events, to multimillion-dollar infrastructure costs of stadiums, transit systems, airports, and convention centres.

Attractions

In general, attractions are needed to entice visitors to an area, making attractions a fundamental element of tourism. Some regions have been fortunate to have been well endowed with natural attractions, such as climate, scenery (Boti Falls and the Umbrella Rock), or flora and fauna (Tafi Monkey Sanctuary) in Ghana, the Niagara Falls in Canada and the Rocky Mountain National Park in Colorado. Whilst these so-called "natural attractions" are site-specific, built attractions can be developed in most areas. Regions that have been less fortunate with respect to natural attractions have had to entice visitors with built attractions such as theme parks (Ballarat), historic sites (Elmina Castle, The forks national historic site), or convention facilities (Akosombo Dam).

Attractions are essential to the tourism industry as they provide the stimulation for many people to travel. Without attractions, tourism would not exist. Swarbrooke (1995) discussed some of the definitions that have been put forward for attractions, highlighting the fact that there had not been any universally accepted definition of a visitor attraction. Swarbrooke then proposed a typology of attractions comprising four types: the natural environment, man-made structures not designed to specifically attract visitors, man-made structures specifically designed to attract visitors, and special events.

There are four main types of attractions, as illustrated in Figure 27.1.

For types one and two, the main aim and challenge of managers are to minimise the impact of visitors on the attraction and preferably also to minimise the number of visitors.

In the case of the third type of attraction, man-made attractions which are purpose-built to attract tourists, the aim is to maximise visitor numbers and income, generally, although these attractions may also have other objectives, such as generating money that can be spent on conservation, education, or brand extension. In relation to type four, events, the key issues depend on the nature of the event. Where it is a traditional event, the challenge is to maintain the spirit of the event while meeting visitors' needs. If it is a new event, the main aim is usually to attract visitors and maximise income.

Figure 27.1 The four types of visitor attractions.

Events Marketing

Events are considered as attractions. The main distinguishing feature of events in relation to the other types of attractions is the fact that events by their very nature are temporary, which is an essential element of their appeal (Getz, 2022). A major benefit of event tourism is that for many events, little additional infrastructure is required and they can be held in most regions. In principle, they can be scheduled at times and in places to reduce the impacts of seasonality or to reduce crowding and damage in more sensitive areas (Getz, 1991; Laing, 2018). "This is the age of events" (Janiskee, 1996, p. 100, Drinkwater et al. (2022), with events satisfying consumers' needs for "structured leisure experiences that are high in entertainment value" (p.100).

Robinson and Noel (1991), Laing (2018), however, make the point that events are in fact an alternative to the highly structured leisure programmes of the past. "Although the majority of events have probably arisen for non-tourist reasons … there is clearly a trend to exploit them for tourism and to create new events deliberately as tourist attractions" (Getz, 1989, p. 125, Getz, 2008). Events can be an important motivator for travel, both day trips and overnight travel. Pleasure travel as a result of attendance at events accounts for about 3 percent of the total pleasure travel in the United States but it is one of the fastest-growing segments of the tourism industry (Backman et al., 1995). In a study conducted by Wicks and Fesenmaier (1995) and Mair and Weber (2019), that involved a survey of 2100 randomly selected households, it was found that 57 percent of all pleasure trips in the previous year had included an event. Of these, 55 percent indicated that attendance at an event resulted in an overnight stay, which demonstrated the importance of the field of events to the tourism industry.

An emerging trend in the tourism and leisure fields is that a growing number of people are tending to seek more participative experiences, hence the increasing use of the term "experiential tourism." These people are no longer satisfied with simply looking at a tourist attraction and wish to be in some way involved with the experience. Consumer satisfaction with a tourist attraction is enhanced with visitor participation (Pearce, 1991, Swarbrooke and Page, 2012). Events, particularly festivals, are important from this perspective as they often provide the attendee with the opportunity for a participative experience. Participation can take many varied forms including the tasting of local produce and the involvement in games and activities aligned with the theme of the event. There are some notable exceptions to this generalisation, such as the Olympic Games, which is highly successful in terms of patronage but does not really offer a great deal in the way of a participative experience.

Events are the new "Image Builders" and "are starting to dominate natural or physical features in the identification of cities" (Burns, Hatch, and Mules, 1986, p. 5, McHone and Rungeling, 2000). They can have very large impacts on a host region and the types of impacts are varied, which means that a given event can be staged for a large number of reasons. The reasons for staging events are found throughout the literature and can be summarised as:

- Increased visitation to a region.
- Positive economic impact.
- Increased employment.
- Improvement of a destination's image or awareness.
- Enhanced tourism development.
- Ability to act as a catalyst for development.
- Reduction of seasonal fluctuations or extension of the tourism season.
- Animation of static attractions.
- Enhanced community pride.

Governments have become interested in events largely because of their ability to attract visitors, and hence visitor spending and their ability to raise the awareness of the host region for future tourism (, Torre and Scarborough (2017). However, creating events for particular destinations requires careful planning and implementation.

A Decision Framework: Event Planning

The planning of an event involves answering certain questions of strategic importance. Answering these questions serve as a guide through the decision framework for event planning. The questions include:

What strategic factors relate to this event? To answer this question, one should take into account the following:

- Mission, goals, and objectives of the event-tourism programme.
- Philosophy, orientation, and attitude of the event's managers.
- Environmental scanning (e.g., noting similar events elsewhere).
- Event-management structure.

- Appraisal of the internal organisation in terms of strengths and weaknesses.
- Time frame for attaining targets.

What is the profile of visitors? The profile should include the following factors:

- Who are the visitors?
- Where do they come from?
- The benefits they seek in attending the event.

What is their expenditure profile? This question focuses on the following factors:

- Expenditure per day.
- Item purchase.
- Foreign-exchange earnings.
- Incremental foreign-exchange earnings.
- Regional foreign-exchange distribution.
- Expenditures by local visitors.
- Expenditures by visitors from elsewhere in the country.
- Expenditures by visitors from overseas.
- Total expenditures from all sources.

What are the economic and social costs and benefits of the event? This question addresses financial, social, and physical factors, including the following:

- Revenue estimates.
- Operating-expenditure estimates.
- Infrastructure-cost estimates.
- Cash-flow analysis.
- Sensitivity analysis to gauge the elasticity of demand to price changes.
- Employment estimates.
- Social-cost estimates.

What is the event's profile? This profile comprises the following issues:

- Event history.
- Event proposal.

- Event sponsor.
- Event-rights holder.
- Governing body.
- Event critical path.
- Nature of event support required.
- Media coverage.

Summary

In this chapter, we discussed various strategies and investments that can be undertaken by firms, governments, or other agencies to improve the attractiveness of particular tourist destinations. We focused on the development of attractions as a whole and events as a particular form of attraction. We then considered a decision framework for planning events.

Activity

1. Define attraction in tourism destination marketing and explain the four main types of attractions in destination marketing.
2. Identify a tourist destination in Africa and employ the event planning decision-making framework above to design a unique event for that tourist destination.
3. How does the event planning decision-making help the firm above to make their sites attractive to clients?

References

Backman, K. F., Backman, S. J., Uysal, M., & Sunshine, K. M. (1995). Event tourism: An examination of motivations and activities. *Festival Management and Event Tourism*, *3*(1), 15–24.

Burns, J. P. A., Hatch, J., & Mules, T. J. (1986). The Adelaide Grand Prix: The impact of a special event.

Cohen, E. (1979). Rethinking the sociology of tourism. *Annals of tourism research*, *6*(1),18–35.

Drinkwater, K., Massullo, B., Dagnall, N., Laythe, B., Boone, J., & Houran, J. (2022). Understanding consumer enchantment via paranormal tourism: Part I—Conceptual review. *Cornell Hospitality Quarterly*, *63*(2), 195–215.

Getz, D. (1989). Special events: Defining the product. *Tourism Management, 10*(2), 125–137.

Getz, D. (1991). *Festivals, special events, and tourism.* Van Nostrand Reinhold.

Getz, D. (2008). Event tourism: Definition, evolution, and research. *Tourism Management, 29*(3), 403–428.

Getz, D. (2022). Event management. In *Encyclopedia of tourism management and marketing* (pp. 144–147). Edward Elgar Publishing.

Gnoth, J., & Anwar, S. A. (2000). New Zealand bets on event tourism. *The Cornell Hotel and Restaurant Administration Quarterly, 41*(4), 72–83.

Janiskee, R. L. (1996). Historic houses and special events. *Annals of Tourism Research, 23*(2), 398–414.

Laing, J. (2018). Festival and event tourism research: Current and future perspectives. *Tourism Management Perspectives, 25*, 165–168.

Mair, J., & Weber, K. (2019). Event and festival research: A review and research directions. *International Journal of Event and Festival Management. 10*(3), 209–216.

McHone, W. W., & Rungeling, B. (2000). Practical issues in measuring the impact of a cultural tourist event in a major tourist destination. *Journal of Travel Research, 38*(3), 300–303.

Pearce, P. L. (1991). Analysing tourist attractions. *Journal of Tourism Studies, 2*(1), 46–55.

Robinson, A., & Noel, J. G. (1991). Research needs for festivals: A management perspective. *Journal of Applied Recreation Research, 16*(1), 78–88.

Swarbrooke, J. (1995). *The development and management of visitor attractions.* Oxford: Butterworth Heinemann.

Swarbrooke, J., & Page, S. J. (2012). *Development and management of visitor attractions.* Routledge.

Torre, A., & Scarborough, H. (2017). Reconsidering the estimation of the economic impact of cultural tourism. *Tourism Management, 59*, 621–629.

Wicks, B. E., & Fesenmaier, D. R. (1995). Market potential for special events: A midwestern case study. *Festival Management and Event Tourism, 3*(1), 25–31.

Chapter 28

Developing Tourism and Hospitality Marketing Plans

Chapter Outline

- Introduction
- Purpose of a Marketing Plan
- Executive Summary
- Corporate Connection
- Environmental Analysis and Forecasting
- Segmentation and Targeting
- Next Year's Objectives and Quotas
- Action Plans: Strategies and Tactics
- Resources
- Marketing Control, Monitoring, and Evaluation
- Important Considerations in Marketing Plan Development
- Benefits of a Well-designed Tourism Marketing Plan
- Sustainability and Digital Considerations
- The Hospitality Marketing Plan

Objectives

At the end of this chapter, the student should be able to:

- Explain the importance and purpose of a marketing plan.
- Know how to develop a marketing plan following the process described in this section.

DOI: 10.4324/9781032688497-28

- Know how to sell a marketing plan and prepare for future marketing plans.

Introduction

The effective management of hospitality and tourism marketing is dependent on the quality of the marketing plan that is being implemented and the proactive measures undertaken to strategically respond to changes in the marketing environment. In this chapter, we will discuss a practical guide to developing a marketing plan. Over three decades ago, tourism scholars were discussing the importance of marketing to tourism success and developing conceptual models to address the issue (Papadopoulos 1989) and with discussions around sustainability, there have been newer discussions around marketing or tourism within sustainability frameworks (Jamrozy 2007).

Research shows that businesses that have a business plan have a greater chance of staying in business than those without a plan. By extension, the marketing department will only remain relevant to the success of the hospitality and tourism business with the formulation of an operational plan. Developing a marketing plan goes hand in hand with developing a business plan, but it is possible to develop a marketing plan as a standalone document.

Successful marketing requires planning and careful execution. When little or no time is devoted to planning, the marketing department tends to operate without direction and is reactive rather than proactive. Even experienced managers sometimes fail to see that this is occurring until it is too late. This may be one of the root causes of high turnover within hospitality, marketing, and sales departments.

Purpose of a Marketing Plan

The marketing plan:

- Allows the organisation to look internally to fully understand the impact and results of past marketing decisions.
- Allows the organisation to look externally to fully understand the market in which it chooses to compete.
- Sets future goals and provides direction for future marketing efforts that everyone in the organisation should understand and support.

- Is a key component in obtaining funding to pursue new initiatives.
- Serves several purposes within any hospitality company.
- Provides a road map for all marketing activities of the firm for the next year.
- Ensures that marketing activities are in agreement with the corporate strategic plan.
- Forces marketing managers to review and think through objectively all steps in the marketing process.
- Assists in the budgeting process to match resources with marketing objectives.
- Creates a process to monitor actual against expected results.

The development of a marketing plan is a rigorous process and cannot be accomplished in a few hours. Instead, it is best to set aside one or more days to develop next year's plan usually out of the normal office environment to avoid interruption of the planning process.

To be effective, a new marketing plan must be written each year but must have the long-term strategic objective of the hospitality or tourism enterprise in mind. Many managers believe that the process of writing a plan is invaluable because it forces those writing it to question, think, and strategise. A plan should be developed with the input and assistance of key members of the marketing department.

A marketing plan should contain the following sections:

1. Executive summary.
2. Corporate connection.
3. Environmental analysis and forecasting.
4. Segmentation and targeting.
5. Next year's objectives and quotas.
6. Action plans: strategies and tactics.
7. Resources.
8. Marketing control, monitoring, and evaluation.

Section I: Executive Summary

The executive summary and a few charts or graphs from the body of the plan may be the only parts ever read by the top management. Consequently, it is of great importance to write this section carefully, with the top management in mind. A few tips may assist in writing the executive summary:

- Write it for top executives.
- Limit the number of pages to between two and four.
- Use short sentences and short paragraphs. Avoid using words that are unlikely to be understood.

Organise the summary as follows: describe next year's objectives in quantitative terms; briefly describe marketing strategies to meet goals and objectives, including a description of target markets; describe expected results by quarter; identify the dollar costs necessary, as well as key resources needed.

Read and reread the executive summary several times. Never write it once and then place it in the plan. Modify and change the summary until it flows well, is easily read, and conveys the central message of the marketing plan.

Section II: Corporate Connection

Relationship to other plans.

The marketing plan must be supportive of the firm's long-term strategic decisions. These might include:

- Corporate goals with respect to profit, growth, market share, etc.
- Desired market share.
- Positioning.
- Vertical or horizontal integration.
- Strategic alliances.
- Product line breadth and depth.
- Marketing-related plans.

Often there will be companies with different departments which may not report to the marketing department but are nonetheless related. These might include sales, advertising, and promotion, public relations/publicity, marketing research, pricing, or customer service. If these departments are not reporting directly to the marketing department, a cooperative effort must be made in order to produce an integrated and comprehensive marketing plan.

Corporate Direction

A good marketing plan must recognise the corporate directional elements and should acknowledge to top management the relevance that these played in the development of next year's plan. By incorporating the mission statement, the corporate philosophy, and the corporate goals will reinforce their importance as well as recognise the efforts of top management.

Section III: Environmental Analysis and Forecasting

Businesses don't operate in isolation. There are factors both within the control and beyond the control of businesses that can affect their successes. These include the state of the economy, the general state of the tourism industry, the value of currencies, international events, what competitors are doing, etc.

A full analysis of your business needs to consider the internal and external factors in terms of strengths, weaknesses, opportunities, and threats. This review, known as a SWOT (Strengths, Weaknesses, Opportunities, and Threats), will provide an overview of areas in need of attention and highlight opportunities that can be developed to expand the business.

External Factors

External factors are factors beyond your control that could affect the business, for example, the economy, technology, trends affecting the tourism industry, legal and political issues, competition, the environment, etc. Major environmental factors include:

- Social factors such as crime, changing demographics, and AIDS will vary in intensity and geographical incidence. The marketer who keeps abreast of these trends can often capitalise on them if recognised and exploited before the competition.
- Political factors such as legislation affecting international politics and trade may be capitalised on as well.
- Economic trends. Unemployment levels and income and investment trends should be considered in the company's strategic planning process.

Internal Factors

Internal factors are those over which you have some control, for example, equipment, financial resources, staff, etc. Depending on the nature of your business, there will be several areas that need to be considered, for example, physical resources—buildings and equipment. Are they being fully utilised? Can they cope with expansion?

Then there are the financial resources of the business. Is your cash flow sufficient? Do you have access to additional funds to finance any growth activities? Is a lack of finance impeding your growth?

Finally, you need to look at the experience and expertise of your staff. Are they motivated? Are they being used to the best of their ability? Is there any training available that could improve the efficiency and service offered by your business? The answers to these questions will assist you to understand what your business can achieve.

S.W.O.T

A common way of summarising strengths, weaknesses, opportunities, and threats is in a table format. Use the table to:

Build on your business strengths whenever possible. They should be exploited and reinforced to give your business an edge over competitors. Only correct weaknesses that are holding the business back—don't address minor areas which have little impact on the business. Maximise your business success by taking advantage of opportunities.

Competitive Analysis

It is unlikely that you will operate a business that does not have competition. You must identify who your competitors are, what their strengths and weaknesses are, how much of the business or market share they have, and how much of the market share you have or want to have.

Competitor analysis should be accomplished by exploring two levels:

- Observable physical properties of the competition.
- Intangibles such as service levels, cleanliness, staff knowledge, and sales department.

There is a need to consider your competitors to determine your competitive advantage. A competitive advantage is the "point of difference" that sets a business apart from its competitors. Every business needs a competitive advantage. It gives you an edge over your competitors and provides a focus for marketing activities. True competitive advantages are factors that are recognised by guests and influence their purchase decisions. You can also use competitor analysis to maintain your advantage and seize opportunities that may arise from the actions of your competitors.

Market Trends

The impact of market trends is particularly critical for marketers and can be found free of charge from several sources including chambers of commerce, visitors' bureaus, universities, and trade associations. Three broad trends that might be included in the marketing plan text are listed. They include visitor trends, competitive trends, and related industry trends.

- *Visitor Trends*: origination areas, stopover sites, visitor demographics, spending habits, length of stay, and so on.
- *Competitive Trends*: numbers, location, type of products offered (e.g., all-suite holds), occupancy levels, average rates, and so on.
- *Related Industry Trends*: interdependence of the members of the hospitality industry upon airline flights, convention centre bookings, new airport construction, and new highways. It is important to study trends for supporting or related industries.

Market Potential

A definition might be the total available demand for hospitality or tourism products within a particular geographic market at a given price. Note a definite distinction between different hospitality or tourism products in any estimates of market potential. The purpose of such estimates is to balance capacity with demand and be able to capitalise on opportunities created by high-demand periods. The process of estimating a property's market demand normally begins by examining the total market for all types of hospitality or tourism, but should then shift to the specific markets of the particular hospitality or tourism destination and its direct competitors. Furthermore, estimates should be expressed as demand estimates at various price points. Never assume market potential is static.

Marketing Research

The need for marketing intelligence is ongoing. Much of the information acquired by marketing research in a current calendar or fiscal year will serve as the basis for developing next year's marketing plan. Marketing research needs can usually be divided into macro-market and micro-market information.

1. Macro-market information addresses the big-picture trends and might include industry trends, social/economic/political trends, competitive information, and industry-wide customer data.
2. Micro-market information is retrieved from sources directly connected with the product or service. Examples include guest history information, new product analysis, and testing, pricing studies, key account information, and advertising/promotion effectiveness.

The information should then be analysed to make appropriate linkages that enable the hotel or destination to accurately measure and anticipate the effects of potential marketing efforts.

Section IV: Segmentation and Targeting

Segmentation Analysis: The heart of any marketing plan is a careful analysis of available market segments and the selection of appropriate target markets. Of the available total market, which potential segments are most appropriate and profitable for this property or company? Effective segmentation analysis is achieved by:

- A clear and thorough understanding of who the company is and what it wishes to be. What are the goals of the company? What are the most effective means necessary to get there?
- A thorough comprehension of who the segments are. Of the available segments, what is the capacity of the property to satisfy them? Are they an appropriate means to the goals of the company?

The concept of target markets is one of the most basic, yet most important aspects of marketing. There is no such thing as the "general public." It is unrealistic to think that you can attract everyone. Defining your target market helps you decide where to commit resources and what kinds of promotional methods and messages to use. Targeting is the careful selection of segments, both existing and potential. Examples of target markets include the mature market (55+ years), women, international markets, and families.

Section V: Next Year's Objectives and Quotas

Objectives

Your marketing objectives should be based on understanding your strengths and weaknesses, and the business environment in which you operate. They should also be linked to your overall business strategy. For example, if your business objectives include increasing visitation by 10 percent over the next year. Your marketing objectives might include targeting a promising or emerging new market segment to help achieve this growth. As with any strategic initiative, a marketing plan should start with objectives. Your marketing objectives will guide your entire marketing initiative and be used for evaluation. Without objectives, you may get off-track and will not know when you have reached your ultimate goal.

Your objectives often focus on your specific target market(s). Objectives must:

- Be measurable in quantitative terms, such as number of visitors, sales volume, and so forth. By having quantitative objectives, you will have a clear target to strive towards and will know when the objective has been achieved.
- Be framed within a specific period.
- Be outcome-based. In other words, what is the result you are looking for?

The process of establishing objectives is not an easy task and should not be accomplished by simply adding a random percentage to last year's objectives. Objectives should be established after carefully considering the areas already discussed including corporate goals, corporate resources, environmental factors, competition, market trends, market potential, available market segments, and possible target markets.

Other sub-objectives may also be established by the marketing department. Again, these should support corporate goals and next year's primary objectives. Each marketing support area needs to be guided by a set of sub-objectives. This includes areas such as advertising, promotion, public relations marketing research, and, of course, sales.

Quotas

Quotas are specific but realistic goals imposed on salespeople. They must be supportive of next year's objectives, obtainable and measurable. Most often they are broken down into small units such as each salesperson's quota for the month or week. Aggregated, they must meet or exceed the company's annual objectives.

Based on next year's objectives quotas should thus be:

- Individualised.
- Realistic and obtainable.
- Broken down into small units, such as each salesperson's quota per week.

Understandable and measurable: for example, quota = $10.000 sales for product line x in week 5. An example of a quota that is not understandable or measurable is "to obtain a 10% increase of market share early in the year."

Section VI: Action Plans: Strategies and Tactics

Strategies and tactics, as stated before, are supportive of objectives. This is where the specific strategies for achieving the company's goals are. These might include target markets, positioning, the marketing mix, and any budgetary expectations. Tactics flow from strategies and may incorporate agencies and persons inside and outside of the company. The marketing plan should detail specific strategies and tactics for each target market, any new products or services, advertising campaigns, and pricing structure changes. Accordingly, each of the objectives and sub-objectives will need to be addressed with such detail.

Salesforce strategies begin with six key points which should be supported by specific sales tactics.

a. Sales strategies:
   - Prevent erosion of key accounts.

- Grow key accounts.
- Grow selected marginal accounts.
- Eliminate selected marginal accounts.
- Retain selected marginal accounts but provide lower-cost sales support.
- Obtain new business from selected prospects.

b. Each general strategy is supported by specific sales tactics:
   (1) *Outside the Company*: Sales blitz of all or targeted accounts and projects, telephone, direct mail, and personal sales calls to selected decision-makers and decision influencers, trade booths at selected travel shows, sales calls, and working with travel intermediaries: tour wholesalers, travel agencies, incentive houses, international sales reps, others. Luncheon for key customers, prospects, or decision influencers, travel missions, and other tactics.
   (2) *Inside the Company*: Training of sales staff, involvement, and support of non-sales personnel, motivational and control programmes, involvement and support of management.

Advertising/Promotion Strategies

These should be established by individuals within the company responsible for these strategies in conjunction with support groups. The director of advertising must work with the advertising agencies; the director of marketing might work directly with consultants to develop the most appropriate strategies. Caution is required in this area to retain authority in deriving and implementing advertising and promotion strategies. Outside groups like advertising agencies may not view the objectives the same way as the client does. The resolution to this problem is to work as a team to develop an advertising/promotion mix that is acceptable and productive to meet the stated objectives.

Pricing Strategies

These remain a function of marketing despite a trend to create separate internal pricing departments such as revenue management or yield management. If it does, indeed, fall under the authority of another department, it is necessary that the marketing department continuously interfaces with an internal pricing department in order to develop mutually profitable strategies.

- Reservations departments should be consulted during price planning; they often have considerable latitude to adjust prices and may account for a significant percentage of sales.
- Sales promotions and advertising to target markets must support pricing decisions.
- Pricing affects every facet of marketing and sales; therefore, it should be reviewed and revised frequently.

Product Strategies

New product and existing product strategies are another function of the marketing department and should also be addressed in the marketing plan. Marketers can be expected to have considerable input when planning dramatic new additions to product lines. Likewise, it is also their responsibility to enhance revenue from existing products and services. Hospitality products are changing rapidly. For example, Las Vegas has been transformed from an adults-only playground to a destination resort for families, in competition with Disney.

Section VII: Resources

Resources are needed to support strategies and meet objectives. Therefore, marketing plans must be written with available resources or those likely to become available, in mind. A common error in writing a marketing plan is to develop strategies that are probably highly workable but for which there is insufficient support. Another error is to assume that top management will not provide additional support regardless of the brilliance of the plan. Marketing managers must be able to convince top management of the necessity of employing the marketing plan. Needed resources include:

- *Personnel*: Generally, the most costly and difficult resource needed to ensure success with marketing/sales strategies is personnel. Personnel is usually the most costly and difficult resource needed to ensure the success of marketing/sales strategies. Issues about personnel must cover any additional personnel needed and some justification for the need, training and recruiting costs, and the type of individuals required for a position.

- Equipment and space may also be required to achieve marketing objectives.
- Other monetary support such as travel expenses and incentive rewards should also be included in the marketing plan's budget.
- Research, consulting, and training costs should be budgeted in the plan as well.
- Miscellaneous costs may be an area that will cover extraneous items such as magazine subscriptions and professional association memberships.

Budgets should be established to reflect projected costs on a weekly, monthly, quarterly, and annual basis to guide the department in its resources and their allocation. It is important to include a contingency figure in your budget. It is not always possible to be aware of every marketing opportunity and worthwhile marketing opportunities can arise at short notice.

Section VIII: Marketing Control, Monitoring, and Evaluation

Marketing is an ongoing process. Plans need to be monitored and reviewed regularly. Evaluating your marketing plan will also help you to prepare a more realistic and achievable plan in the future. This section of your plan should also include plans and procedures for tracking each type of marketing strategy and tactic you are using. Tracking helps monitor the effectiveness of each ongoing marketing activity and is especially helpful with your overall programme evaluation at the end of implementation. Important areas for monitoring and evaluation include:

- Sales forecasts and quotas must be continuously evaluated against actual results. This helps to aid in developing corrective tactics to adjust for trends. Failure to keep abreast of them may result in missing opportunities to adjust the tactics or enhance the sales goals.
- Expenditures against Budget: Similarly, actual expenditures must be continuously compared against budget projections set in the marketing plan.
- Periodic evaluation of all marketing objectives by sales and marketing managers is a necessary measure—a critical role is to ensure that all

objectives are met or exceeded promptly. A marketing activity timetable is a device often used by sales departments that list major activities, completion dates, individuals responsible, and space for checking off the task when completed.

- Marketing Activity Timetable: One method commonly used by marketing/sales managers to ensure that tasks are completed on time is the use of a marketing activity timetable. This simple device lists major activities, the dates they must be completed, the person responsible, and space for checking whether the task has been accomplished.

In certain cases, there may be the need to make readjustments to the marketing plan. Market conditions change, disasters occur, and many other reasons create a need to refine marketing plans. Generally, refinements should be made in the area of tactics, budgets, and timing of events rather than in major objectives or strategies.

Important Considerations in Marketing Plan Development

Presenting and selling the plan:

Although the marketing plan is important in effective marketing management there is a need to generate support and understanding across the total operations of the company. Never assume that a marketing plan is so logical that it will sell itself. A marketing plan must be sold to many people. These include the following:

A. *Members of the Marketing/Sales Department*: Often this group is unaware of the benefits that a marketing plan will produce. Managers of these departments need the support of these team members in the planning process; it is, therefore, best to sell the benefits of the process rather than forcing acquiescence.

B. *Vendors/Advertising Agencies and Other Distribution Intermediaries*: Their participation is necessary for the success of the plan and should be included in the planning process.

C. *Top Management*: This group must approve the annual marketing plan. Often this is done at a luncheon or formal presentation generating great excitement within the sales/marketing department. The presentation should be professional, planned, and rehearsed. It is recommended that various media be used; colour charts and computerised presentations are examples of impressive materials. This step should not be underemphasised.

Preparing for the Future

The process of marketing planning is a continuum. The task is never-ending. Marketing/sales managers must always be planning. In reality, the development of next year's marketing plan begins the day this year's plan is approved. Marketing plan development depends on the availability of reliable information. This task can always be improved. The process of data collection and analysis from internal and external sources continues each day. Marketing/sales managers must always be alert for methods to improve the process.

Sustainability and Digital Considerations

Creating a successful tourism marketing plan requires careful planning, research, and a well-thought-out strategy. In the light of the increasing importance of sustainability issues and digital management, the chapter closes with the outline of tourism marketing plan for a National Tourism Authority that considers these two key imperatives:

1. Market Research
   - *Target Audience*: Identify your target demographic, including age, gender, interests, and preferences.
   - *Market Trends*: Research current trends in the tourism industry, such as emerging destinations, traveller behaviours, and technology advancements.
   - *Competitor Analysis*: Analyse your competitors' marketing strategies, pricing, and offerings.
2. Setting Objectives
   - Define clear and measurable objectives for your tourism marketing plan. Common objectives include increasing tourist arrivals, boosting revenue, or enhancing brand awareness.
3. SWOT Analysis
   - Conduct a SWOT analysis (Strengths, Weaknesses, Opportunities, Threats) to assess your destination's internal and external factors.

4. Unique Selling Proposition (USP)
   - Determine what makes your destination unique and why travellers should choose it over others. Highlight these features in your marketing materials.
5. Marketing Mix (4Ps)
   - *Product*: Define your tourism product, including accommodations, activities, and experiences.
   - *Price*: Set competitive pricing strategies that align with your target audience.
   - *Place*: Determine distribution channels, such as travel agencies, online booking platforms, and partnerships.
   - *Promotion*: Develop your marketing and advertising strategies.
6. Content Creation
   - Create engaging and high-quality content, including photos, videos, blog posts, and social media updates, to showcase your destination.
7. Digital Marketing
   - Utilise various online marketing channels, such as social media, search engine optimisation (SEO), email marketing, and pay-per-click (PPC) advertising, to reach your target audience.
8. Website Development
   - Ensure your destination has an attractive and user-friendly website with essential information, booking capabilities, and an easy-to-navigate layout.
9. Social Media Engagement
   - Actively engage with your audience on social media platforms by sharing updates, responding to comments, and running targeted campaigns.
10. Partnerships and Collaborations: Collaborate with airlines, hotels, travel agencies, and local businesses to create package deals and cross-promotional opportunities.
11. Public Relations: Work on building relationships with travel journalists, bloggers, and influencers who can help promote your destination through reviews and features.
12. Advertising Campaigns: Develop advertising campaigns tailored to your target audience, utilising both online and offline media.
13. Monitoring and Measurement: Implement tracking tools to monitor the effectiveness of your marketing efforts, such as website analytics, social media insights, and customer feedback.

14. Budgeting: Allocate a budget for each marketing initiative and monitor expenses to ensure you stay within your financial constraints.
15. Evaluation and Adjustment: Regularly evaluate your tourism marketing plan's performance and make necessary adjustments based on data and feedback.
16. Crisis Management: Develop a crisis management plan to address unforeseen challenges, such as natural disasters or negative publicity.
17. Sustainability: Promote sustainable tourism practices and eco-friendly initiatives to attract responsible travellers.

Remember that a successful tourism marketing plan is an ongoing process. Continuously adapt and refine your strategies based on changing market conditions, traveller preferences, and emerging trends in the tourism industry. Regularly measure your results and make data-driven decisions to achieve your objectives and stay competitive.

Benefits of a Well-Designed Tourism Marketing Plan

The benefits of a well-designed tourism marketing plan are outlined next:

A well-executed tourism marketing plan offers numerous benefits to destinations, tourism businesses, and local communities. Here are some of the key advantages:

1. *Increased Tourist Arrivals*: Effective marketing strategies can attract more tourists to the destination, resulting in higher occupancy rates for hotels, increased visitation to attractions, and more revenue for local businesses.
2. *Revenue Generation*: A successful marketing plan can lead to increased tourist spending on accommodations, food, activities, and souvenirs, thereby boosting the local economy.
3. *Brand Awareness*: Consistent marketing efforts can enhance the destination's brand recognition, making it more appealing and memorable to potential travellers.
4. *Competitive Advantage*: A well-defined marketing plan can help a destination stand out in a crowded tourism market by highlighting its unique attractions, experiences, and offerings.
5. *Diversification of Tourism Markets*: By targeting specific demographics and markets, a marketing plan can reduce a destination's reliance

on a single market segment, making it more resilient to economic fluctuations.

6. *Job Creation*: Increased tourism can lead to the creation of jobs in various sectors, including hospitality, transportation, food services, and entertainment, benefiting local communities.
7. *Infrastructure Development*: Tourism revenue can be reinvested in infrastructure improvements, such as better transportation, upgraded facilities, and environmental conservation efforts.
8. *Cultural Exchange*: Tourism can facilitate cultural exchange and understanding as visitors interact with local communities, learn about different cultures, and share their own experiences.
9. *Community Engagement*: A well-planned marketing strategy can involve local communities, encouraging their active participation in promoting tourism and fostering a sense of pride in their region.
10. *Environmental Conservation*: Sustainable tourism marketing plans can promote responsible travel practices, reduce the environmental impact of tourism, and support conservation efforts.
11. *Seasonal Balancing*: Effective marketing can help distribute tourist arrivals more evenly throughout the year, reducing the strain on resources during peak seasons.
12. *Destination Improvement*: Tourism revenue can be reinvested in destination improvements, enhancing the overall visitor experience and making the destination more appealing.
13. *Data-Driven Decision-Making*: Marketing plans rely on data and analytics, allowing destinations to make informed decisions and adjust strategies based on real-time feedback and trends.
14. *Partnerships and Collaborations*: Collaborations with airlines, travel agencies, and other businesses can result in cost-effective marketing initiatives and package deals that benefit all stakeholders.
15. *Long-term Sustainability*: Sustainable tourism marketing plans can encourage environmentally responsible practices, protecting the destination"s natural and cultural assets for future generations.
16. *Positive Local Perceptions*: Tourism marketing can help combat negative stereotypes and perceptions about a destination, leading to increased local support for tourism.
17. *Crisis Preparedness*: A well-prepared marketing plan includes crisis management strategies, allowing destinations to respond effectively to unexpected events and minimise reputational damage.

In summary, a tourism marketing plan can have a profound impact on a destination's economic, social, and environmental well-being. When executed thoughtfully and strategically, it can lead to increased tourism revenue, job creation, improved infrastructure, and a stronger sense of community engagement, all while preserving the destination's unique character and assets.

The Hospitality Marketing Plan

Creating a comprehensive hospitality marketing plan is essential for hotels, resorts, restaurants, and other businesses in the hospitality industry to attract guests, increase bookings, and build brand loyalty. Below is a framework for a hospitality marketing plan:

1. Executive Summary
   - Provide a concise overview of the entire marketing plan, including goals, target audience, key strategies, and expected outcomes.
2. Situation Analysis
   - *Market Analysis*: Describe the current state of the hospitality market, including trends, demand, and competitive landscape.
   - *SWOT Analysis*: Identify the strengths, weaknesses, opportunities, and threats related to your business.
   - *Customer Analysis*: Profile your ideal guests, including demographics, preferences, and behaviours.
3. Marketing Objectives
   - Clearly define SMART (Specific, Measurable, Achievable, Relevant, Time-bound) marketing objectives, such as increasing room bookings by a certain percentage or achieving a certain level of revenue.
4. Target Audience
   - Specify your primary and secondary target audiences. Consider factors like age, gender, income, interests, and geography.
5. Branding and Positioning
   - Describe your brand identity, values, and unique selling points that differentiate your business from competitors.
   - Explain how you want to position your hospitality business in the minds of your target audience.
6. Marketing Strategies
   - *Product Strategy*: Highlight the unique features and amenities your hospitality business offers.

- *Pricing Strategy*: Determine pricing strategies, including dynamic pricing, seasonal pricing, and discounts.
- *Distribution Strategy*: Outline how and where guests can book accommodations or make reservations (e.g., website, OTAs, direct bookings).
- *Promotion Strategy*: Detail your advertising, content, and promotional efforts.
- *Partnerships and Collaborations*: Identify potential partnerships with local businesses, influencers, or travel agencies to expand your reach.

7. Online Presence
   - *Website*: Ensure your website is user-friendly, mobile-responsive, and optimised for search engines.
   - *Social Media*: Develop a content calendar for social media platforms and engage with your audience regularly.
   - *Online Reviews and Reputation Management*: Monitor and manage online reviews on platforms like TripAdvisor, Yelp, and Google My Business.
8. Content Marketing
   - Create valuable and engaging content, including blog posts, videos, and interactive guides, to showcase your property and local attractions.
9. Email Marketing
   - Build and segment your email list to send personalised offers, newsletters, and updates to past and potential guests.
10. Advertising and Promotion: Implement paid advertising campaigns, including PPC (Pay-Per-Click), social media ads, and display advertising, targeting your ideal audience.
11. Public Relations: Develop relationships with travel journalists, bloggers, and influencers to secure media coverage and endorsements.
12. Events and Promotions: Plan special events, packages, and promotions to attract guests during specific seasons or occasions.
13. Monitoring and Analytics: Set up tracking tools to measure the success of your marketing efforts, including website analytics, conversion rates, and ROI (Return on Investment).
14. Budget: Allocate a budget for each marketing initiative and regularly review spending to stay within budget.
15. Timeline: Create a timeline with milestones and deadlines for implementing various marketing strategies and campaigns.

16. Evaluation and Adjustment: Continuously evaluate the performance of your marketing plan and make necessary adjustments based on data and feedback.
17. Training and Staff Involvement: Train staff on the marketing plan's key elements and involve them in promoting the business.

Remember that a successful hospitality marketing plan is dynamic and adaptable. H&T Planners and Executives must regularly review and update their strategies to align with changing market conditions and guest preferences, ensuring their businesses remains competitive and appealing to their target audiences.

Activity

1. Discuss any five benefits of designing an effective tourism marketing plan.
2. Outline and explain the steps for designing an effective marketing plan for the tourism industry.
3. Outline and explain the steps for designing an effective marketing plan for the hospitality industry.

Summary

In this chapter, we discussed various strategies and investments that can be undertaken to deliver solid tourism and hospitality marketing plans and the advantages of developing these.

References

Jamrozy, U. (2007) Marketing of tourism: A paradigm shift toward sustainability, *International Journal of Culture, Tourism and Hospitality Research*, Vol. 1, No. 2, pp. 117–130.

Papadopoulos, S. I. (1989) A conceptual tourism marketing planning model: Part 1, *European Journal of Marketing*, Vol. 23, No. 1, pp. 31–40.

Index

C

D

E

I

J

K

L

M

Q

R

T